AF710993

KAUAI'S GEOLOGIC HISTORY

A SIMPLIFIED GUIDE

Chuck Blay

Robert Siemers

TEOK
The Edge Of Kauai
INVESTIGATIONS

INTRODUCTION

Kauai and the rest of the Hawaiian Islands are geologically, geographically and biologically unique. **The 132 islands, reefs and shoals** that comprise the **State of Hawaii** extend **over 1500 miles** from the "Big Island" of Hawaii on the southwest to Kure Atoll at the northwestern end of the archipelago. These dots of land are the **Earth's most isolated islands**, lying in the middle of the North Pacific Ocean some 2400 miles from both the nearest continental land mass, North America, and other islands of Polynesia in the South Pacific Ocean. **The islands of Hawaii were one of the last places on Earth occupied by humans.** There is little or no direct evidence of human contact of any kind before about 100 AD. The first significant colonies, made by ocean voyaging Polynesians, were not established until around 400-600 AD. Such was due to Hawaii's geographic location, itself a direct result of its geologic uniqueness.

Hawaii is a place where **the main elements of the Earth's surface and subsurface** interact in a dynamic, and often dramatic, fashion. As new earth has been created by the volcanic production of the Hawaiian mountain-islands above the floor of the Pacific Ocean, and eventually above sea level, the **lithosphere** (land) has been brought into contact with the **hydrosphere** (ocean) and **atmosphere** (air). Subsequently, prior to the appearance of humans, a unique portion of the Earth's **biosphere** developed as plants and animals migrated to and evolved on these severely isolated islands. Over 4,300 species of Hawaii's land plants and animals (including 2,300 insects and 1,000 land snails) exist only in the eight high Hawaiian Islands. Scientists refer to such species as "endemic". The parent species from which these endemic species evolved somehow migrated to the present islands, or to pre-existing islands of the Hawaiian-Emperor Volcanic Chain, across more than 2,000 miles of open ocean, many with the help of birds. Not surprisingly, no flightless land mammals are endemic to Hawaii. A single species of bat and the monk seal, which do spend some of their time on land, are the only known mammals. Again, the geologic/geographic setting has been the ultimate controlling element.

This booklet deals mainly with aspects of that portion of the Earth's lithosphere occupied by the Hawaiian Islands, and even more specifically with the geologic history of the Island of Kauai. The book is organized loosely into four related portions. By design, only a small amount of text is provided. Most illustrations are modifications of key technical diagrams derived from the geologic literature. The captions have been simplified and expanded to enhance their impact. The first portion of the book (illustrations on pages 5-16) provides an overview of the character and origin of the greater Hawaiian mountain-island system, which has been produced over the past 75-80 million years as the Pacific Tectonic Plate moved over the Hawaiian Magmatic Hot Spot. A second portion (illustrations on pages 17-28) deals specifically with Kauai's geologic history. That is followed, on pages 29-33, by key illustrations, never before presented in a simplified book form, that summarize new data on the geology of the submarine portions of the volcanic mountains that support the Hawaiian Islands. A final illustration, the Kauai/Earth Time Line (page 35), and the Concluding Remarks present various aspects of Kauai, and the Hawaiian Islands in general, within a framework of the 10 orders of magnitude of geologic time.

KEY POINTS

The following is a list of key points illustrated by this book.

1. The eight main **high Islands** of the Hawaiian Archipelago include Hawaii, Maui, Oahu, Kauai, Molokai, Lanai, Niihau and Kahoolawe (listed in order of size). They occupy, at the southeast end, only about **350 miles** of the 1525 mile long island chain, and range in geologic age from **less than 0.5 million years** (oldest portion of the island of Hawaii) successively to only slightly **greater than 5.0 million years** (Kauai and Niihau). These eight islands make up over 99 percent of the 6,425 square miles of land area of the state of Hawaii. They include some of the **Earth's largest mountains**, rising from oceanic depths of greater than 18,000 feet to a height above sea level of nearly 14,000 feet, in the case of Mauna Kea on the Big Island of Hawaii. That's a mountain with relief of 32,000 feet.

2. The entire **Hawaiian-Emperor Volcanic Chain**, which extends for **3105 miles** across the North Pacific Ocean, actually comprises more than **107 individual volcanic mountains**. The tops of most no longer rise above sea level to form islands. They range from the eight high volcanic islands (which comprise portions of at least 15 of those volcanic mountains), to the northwestern low volcanic islands and coral atoll type islands, and to an extensive chain of submerged oceanic mountains (now called **"seamounts"**). Kauai, the oldest of the high islands, is only slightly greater than 5.0 million years old. The low islands to the northwest, 480 to 1525 miles away from the Kilauea volcanic crater on the Big Island, are successively older to around **28 million years at Midway**. The submerged **Emperor Seamounts** extend on as far north as the North Pacific Aleutian Submarine Trench where they are greater than **70 million years old**.

3. Based on extensive geologic dating it has been determined that the **average rate of propagation of the Hawaiian-Emperor Volcanic Chain has been 3.4 inches/year** (8.6 cm/yr). That is, the volcanic chain has been formed by the growth of volcanic ocean mountains as the Earth's **Pacific Tectonic Plate** moved across **the Hawaiian Magmatic Hot Spot** at a rate of three and a half inches per year. The plate is still moving at that rate.

4. About **40-43 million years ago, a change in the direction of movement** of the Pacific Tectonic Plate, from northward to northwestward, resulted in the **dog leg bend** in the Hawaiian-Emperor Volcanic Chain. That change is thought to be related to the separation of the Australian and Indian tectonic plates from Antarctica, subsequent movement northward and the eventual **collision of the Indian Plate with Asia** to form the Himalayan Mountains.

5. Most of the volcanic mountain-island systems of the Hawaiian-Emperor Chain have evolved through several **discernible stages of growth, emergence, decline and submergence**. Volcanic stages of eruption, and growth include; **a) shield building**, **b) post-shield collapse and renewed eruption**, and **c) rejuvenation stage eruption**, each with the production of distinctive types of volcanic rocks.

6. The volcanic mountain-islands have undergone extensive destruction by **large scale subsidence, structural failure** (e.g., central collapse and marginal slumps) and **mass wasting** of submarine slopes, including giant landslides that extend out away from many of the islands for up to over 100 miles. The spectacular sea cliffs along the northern portion of Molokai and the Napali coast of Kauai represent giant fault scarps resulting from such structural failures. Extensive **atmospheric erosion**, which has formed deep river valleys, and moderate to extensive **oceanic shoreline erosion** also have helped reduce the size of the islands by the removal of sedimentary products to the surrounding deep ocean floor.

7. Kauai, the oldest of the eight major high Hawaiian Islands, displays most of the products of the **constructive and destructive processes** operating on these mid-ocean volcanic mountain-islands. The four volcanic rock members of the Waimea Canyon Basalt represent the bulk of the island. The shield-building phase is represented mainly by the **Napali Member** (age 5.10-4.35 my), which is particularly well displayed in the Napali sea cliffs. The post-shield phase, which began with massive collapse along the Waimea Canyon fault scarp, included eruption of the apparent caldera-filling **Olokele Member** (age to 3.95 my) and the graben-filling **Makaweli Member** (age 4.16-3.92 my). The **Haupu Member** (precise age unknown) apparently filled a small flank caldera on the southeastern portion of the present island. Rejuvenation stage volcanism resulted in the production of lavas and explosive volcanic vents of the **Koloa Volcanics** (age 3.65-0.52 my) which cover, like a thin veneer, most of the eastern half of the island.

8. Until quite recently Kauai was thought to have been created as one large shield volcano; however, recent mapping of the submarine portion of the mountain-island and investigation of the magneto-stratigraphy of the Waimea Canyon Basalt units indicate that the present island of **Kauai may actually be composed of at least two shield volcanoes.** Detailed mapping of the seafloor also has better demonstrated that Kauai and Niihau once, as recently as one million years ago, formed a single, much larger island.

9. The production of carbonate sediments (i.e., calcium-carbonate skeletal deposits), mainly within the fringing **coral/algal reef** systems and other shoreline sedimentary environments such as the **beaches**, continues to add to the island of Kauai. However, continued **slow subsidence** of the mountain-island and destructive **erosional processes** within the atmosphere and hydrosphere are destroying the island at a significant rate, geologically speaking. Within another 2 to 3 million years Kauai will be reduced to nothing more than a low island, similar to the islands of Nihoa or Necker, incapable of supporting a diversified flora and fauna.

10. All **geomorphic features** and local climatic aspects of Kauai can be explained in terms of the geologic history of the island. Examples include; **a)** the fault scarp that initiated the erosion of **Waimea Canyon**, **b)** the fault scarp resulting in the spectacular sea cliffs of the **Napali Coast**, **c)** collapse of the eastern portion of the island to produce the **Lihue Basin**, and **d)** the eastern-facing fault scarp of **Mt. Waialeale**, across which dramatic orographic lifting of warm, moist, tradewind-driven air results in one of the rainiest places on Earth.

Hawaiian Archipelago

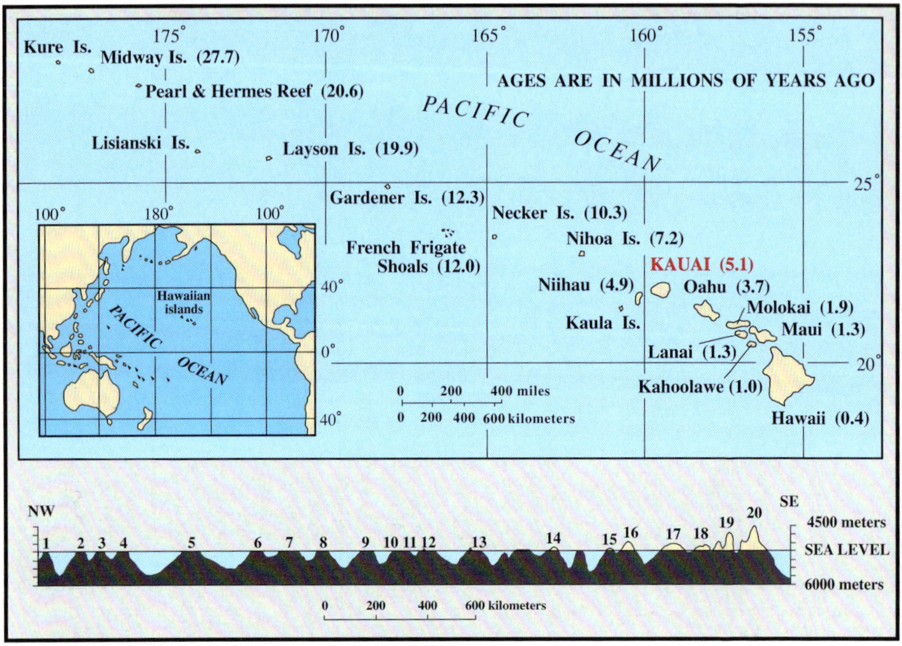

Map and profile of major high islands and related low island features of the **Hawaiian Volcanic Island Archipelago** emphasizing the submerged and eroded character of the islands northwest of Kauai, many of which are capped by reef deposits to form various atoll type islands. These volcanic mountain/islands start as much as 18,000 feet below sea level and extend to as much as 14,000 feet above sea level. Numbers in parentheses indicate average absolute ages (in millions of years ago) as determined by radiometric dating techniques. Features indicated by numbers in the lower profile view are as follows: 1) Kure Island, 2) Midway Islands, 3) Gambia Shoal, 4) Pearl and Hermes Reef, 5) Fisher Reef and Neva Shoal, 6) Laysan Island, 7) Maro Reef, 8) Raita Bank, 9) Gardner Island, 10) St. Rogatien Bank, 11) Brooks Bank, 12) French Frigate Shoals and La Perouse Rocks, 13) Necker Island, 14) Nihoa Island, 15) Niihau, 16) Kauai, 17) Oahu, 18) Molokai, 19) Maui, 20) Hawaii. [modified from Macdonald, et al. (1983), as modified from Stearns (1946)]

The 132 islands that make up the **state of Hawaii** stretch across the central North Pacific Ocean for approximately 1600 miles (2560 km), equivalent to the distance from western Louisiana to San Francisco. The **eight high islands** at the southeastern end of the chain span nearly 400 miles, but are separated by approximately 2100 statute miles (3400 km) from any other sizable land mass.

Hawaiian-Emperor Volcanic Chain

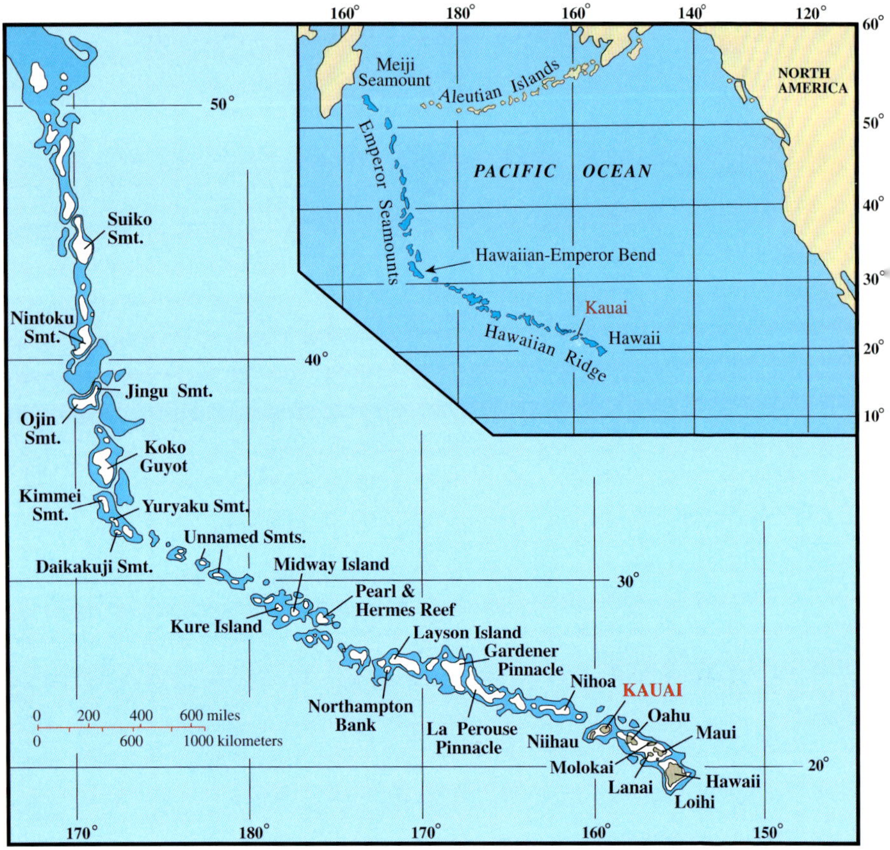

General bathymetry of the **Hawaiian-Emperor Volcanic Chain** which formed as the **Pacific Tectonic Plate** moved northward and then northwestward across the **Hawaiian Magmatic Hot Spot** in the central North Pacific Ocean. More than **107 individual volcanoes** have been delineated within the **5750 Km** (3105 mi) long chain; the Hawaiian Island/Seamount Chain is 3450 km (1860 mi) and the Emperor Seamount Chain is 2300 km (1245 mi). Island/seamount systems are outlined by 1- and 2-km depths in the area of the linear volcanic chains; average depth of the ocean floor in the Hawaiian region is 5 km (3 mi). Significant **absolute ages** of selected islands (in millions of years) include: island of Hawaii, 0-0.4; **Kauai, 5.1**; Midway Islands (including Kure Is.), 27.7; Kimmel and Yuryaku Seamounts (at elbow area of dog leg), 39.9-43.4; Suiko Seamount, 64.7; Meiji Seamount, approx. 75-80. The **change in the direction** of movement of the Pacific Tectonic Plate about **40-43 million years ago** is thought to be related to the separation of Australia from Antarctica and the collision of India with Asia to produce the Himalayan Mountains. (modified from Clague and Dalrymple, 1987 and 1989)

The **Hawaiian/Emperor Volcanic Chain** covers a significant portion of the Earth's surface. When superimposed over the area of North America it extends northwestward from Miami to near Reno and then north through Seattle to the panhandle of Alaska.

Table summarizing **absolute geologic ages** of islands and seamounts along the **Hawaiian-Emperor Volcanic Chain**, based on Potassium-Argon (K-Ar) radiometric determinations. (simplified from Clague and Dalrymple, 1989)

Volcano No.	Name	Distance from Kilauea: Kilometers	(Miles)	Radiometric Age (Millions of Years)
High Islands:				
1	Kilauea	0		0 - 0.4
3	Mauna Kea	54	(34)	0.375
5	Kohala	100	(62)	0.43
6	East Maui	182	(113)	0.75
7	Kahoolawe	185	(115)	1.03
8	West Maui	221	(137)	1.32
9	Lanai	226	(140)	1.28
10	East Molokai	256	(159)	1.76
11	West Molokai	280	(174)	1.90
12	Koolau	339	(210)	2.6
13	Waianae	374	(232)	3.7
14	Kauai	519	(322)	5.1
15	Niihau	565	(351)	4.9
Leeward Low Islands:				
17	Nihoa	780	(484)	7.2
20	unnamed	913	(567)	9.6
23	Necker	1058	(657)	10.3
26	La Perouse Pinnacle (French Frigates Shoal)	1209	(751)	12.0
27	Brooks Bank	1256	(800)	13.0
30	Gardner Pinnacles	1435	(891)	12.3
36	Laysan	1818	(1129)	19.9
37	Northampton Bank	1841	(1143)	26.6
50	Pearl & Hermes Reef	2291	(1423)	20.6
52	Midway Islands	2432	(1510)	27.7
Seamounts:				
57	unnamed	2600	(1615)	28.0
63	unnamed	2825	(1754)	27.4
65	Colahan	3128	(1942)	38.6
65a	Abbott	3280	(2037)	38.7
67	Daikakuji	3493	(2169)	42.4
69	Yuryaku	3520	(2186)	43.4
72	Kimmei	3668	(2285)	39.9
74	Koko	3758	(2334)	48.1
81	Ojin	4102	(2547)	55.2
83	Jingu	4175	(2593)	55.4
86	Nintoku	4452	(2765)	56.2
90	Suiko	4794	(2977)	59.6
91	Suiko	4860	(3018)	64.7

Cross-Plot of Geologic Ages of the Hawaiian-Emperor Volcanic Chain

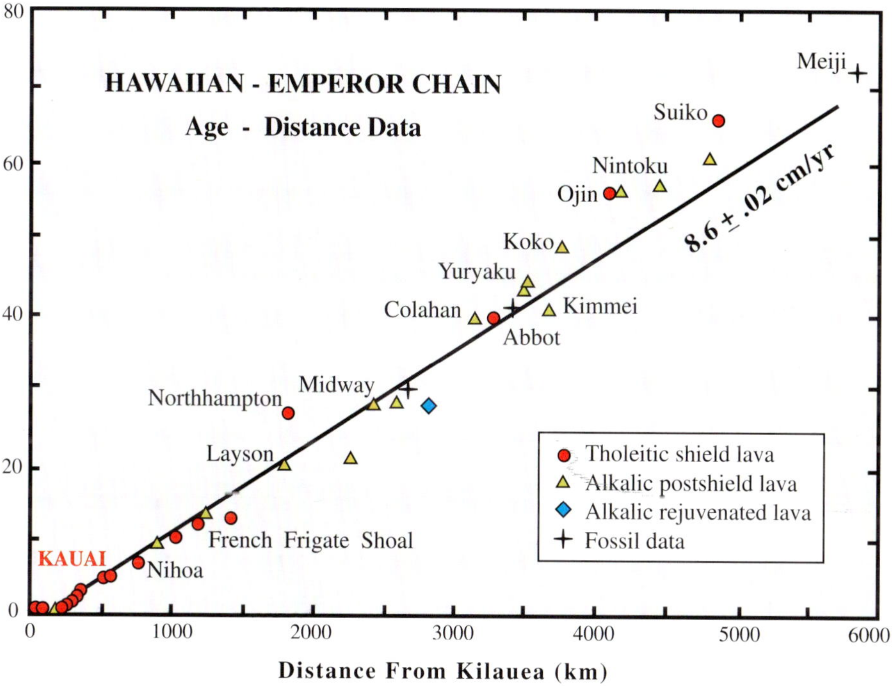

Cross-plot of absolute geologic age of volcanoes in the Hawaiian-Emperor Volcanic chain and their distances from the present-day active volcano of Kilauea on the island of Hawaii. The least-squares fit regression of the data (solid line) indicates that the average **rate of propagation** of volcanism (i.e., **rate of movement of Pacific Tectonic Plate across the Hawaiian Magmatic Hot Spot**) for the entire Hawaiian-Emperor chain was about **8.6 cm/yr** (3.4 inch/yr). Individual shorter segments of the chain indicate slightly slower or faster rates of propagation, ranging from about 6.8 to 11.3 cm/yr (2.7-4.5 inch/yr). (modified from Clague and Dalrymple, 1989)

Hawaiian Mountain/Island Heights

Table summarizing **present heights** (or depths below sea level), **previous maximum heights**, and **absolute geologic ages** of the present major **Hawaiian volcanoes**. Note that Kauai has subsided (sunk) at least 1000 meters (over 3000 ft.) since its volcanic origin over 5 million years ago.(modified from Carson and Clague, 1995)

Volcano	Heights: Present		Heights: Maximum		Age
	Meters	**(Feet)**	**Meters**	**(Feet)**	**(millions of years)**
Lo'ihi	- 950	(-3,116)	- 950	(-3,116)	no data
Kilauea	1,247	(4,090)	1,247	(4,090)	0-0.4
Mauna Loa	4,169	(13,674)	4,169	(13,674)	0-0.4
Mauna Kea	4,205	(13,792)	4,600	(15,088)	0.38
Hualalai	2,521	(8,269)	2,950	(9,676)	no data
Kohala	1,670	(5,478)	2,670	(8,758)	0.43
Mahukona	-1,100	(-3,608)	235	(771)	no data
Haleakala	3,055	(10,020)	5,000	(16,400)	0.75
Kaho'olawe	450	(1,476)	2,100	(6,888)	1.03
West Maui	1,764	(5,786)	3,300	(10,824)	1.76
Lana'i	1,027	(3,369)	2,200	(7,216)	1.28
East Moloka'i	1,515	(4,969)	3,300	(10,824)	1.76
West Moloka'i	421	(1,381)	1,600	(5,248)	1.9
Penquin Bank	- 200	(- 656)	1,000	(3,280)	no data
Ko'olau	960	(3,149)	1,900	(6,232)	2.6
Wai'anae	1,231	(4,038)	2,200	(7,216)	3.7
Kaua'i	1,598	(5,241)	2,600	(8,528)	5.1
Ni'ihau	390	(1,279)	1,400	(4,592)	4.9
Ka'ula	168	(551)	800	(2,624)	4.0
Nihoa	277	(909)	1,300	(4,264)	7.2
Necker	84	(275)	1,100	(3,608)	10.3

Earthquakes in Hawaii (1962-1980, Magnitude > 3.0)

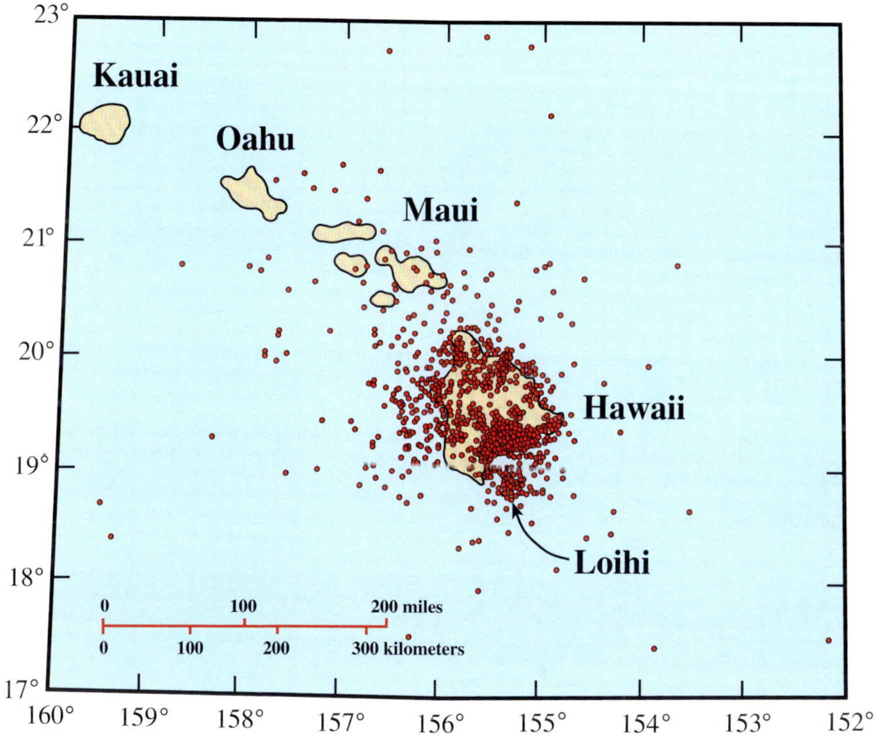

Earthquakes in Hawaii result mainly from the injection of molten magma from the upper mantle upward through the brittle crust. A plot of all earthquakes with magnitude greater than 3.0 (on the Richter scale) for the years 1962-1984 reveal a **general concentration of seismic activity around the island of Hawaii**, and especially in the vicinity of the Kilauea volcanic vent. Such a distribution is indicative of the magmatic "hot spot" below the big island. And now, a new submarine volcanic vent, Loihi, is growing from a present depth of about 2000 ft just to the southeast. Loihi is just more evidence of the continued **northwestward movement of the Pacific Tectonic Plate across the Hawaiian Hot Spot**. (modified from Klein and Koyanagi, 1989)

Tectonic Movement of Hawaiian Islands

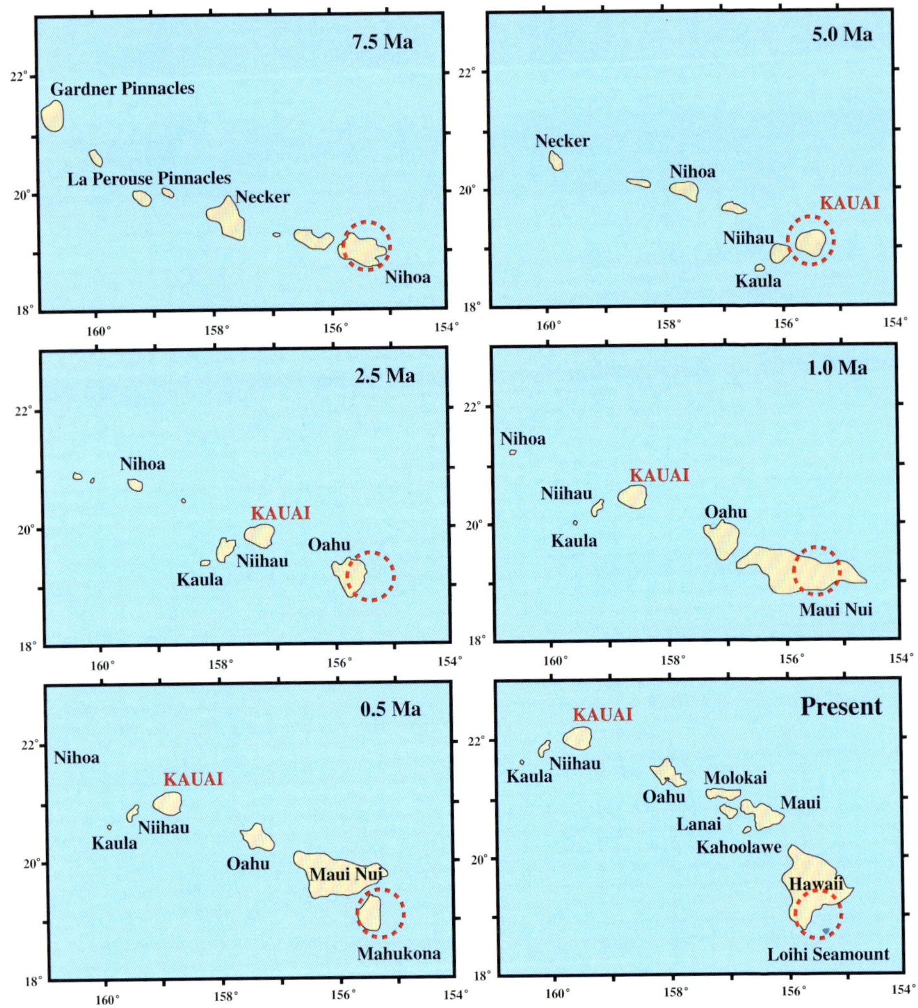

Maps showing coordinate positions in the Pacific Ocean of the present **Hawaiian Islands** and their **reconstructed positions** at five different times (in millions of years), as affected by movement of the Pacific Tectonic Plate across the **Hawaiian Magmatic Hot Spot**. The **rate of movement** of the plate was assumed to be **9 cm/yr** (3.5 inch/yr). In each map, the hot spot (located at 19° N & 155.5° W) is indicated by the red dashed circle. The sizes and fused state of islands, ridges and coral reefs (brown) are estimated based on bathymetry and GLORIA side-scan sonar data. (modified from Carson and Clague, 1995)

Tectonic Plate Model
Hawaiian-Emperor Volcanic Chain

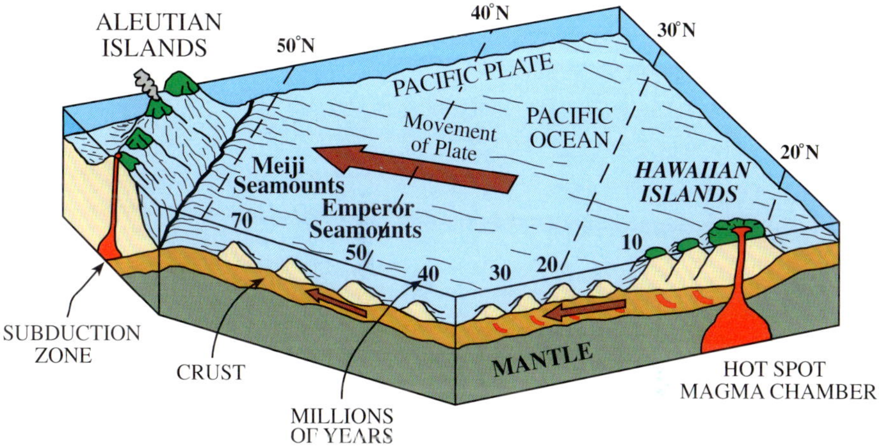

Schematic block diagram of **Hawaiian-Emperor Volcanic Chain**. Volcanic mountain/islands are produced at the **magmatic hot spot** as partial **melting** of upper mantle rocks at **60-170 km** (35-105 mi) depth produces molten magma which is in turn forced upward through the more rigid crust. The actual mechanism of magma formation is not well understood, but the **10-15% volume increase** associated with the transformation of crystalline rock to molten magma and **bouyancy of the less dense magma** are the **main forces** driving the magma upward. As the Pacific Tectonic Plate moves across the hot spot a chain of oceanic volcanic mountains is produced. Eventually the mountain/islands subside and are eroded to produce, successively, low islands, reef shoals/atolls, and submerged seamounts. Such a process has been going on in the North Pacific for **more than 70 million years**. The numbers above the peaks indicate ages in millions of years. The oldest seamounts have already been subducted below the Aleutian Islands. (modified from Bradshaw and Weaver, 1993)

Geologic Stages
Ocean Volcanic Mountain/Islands

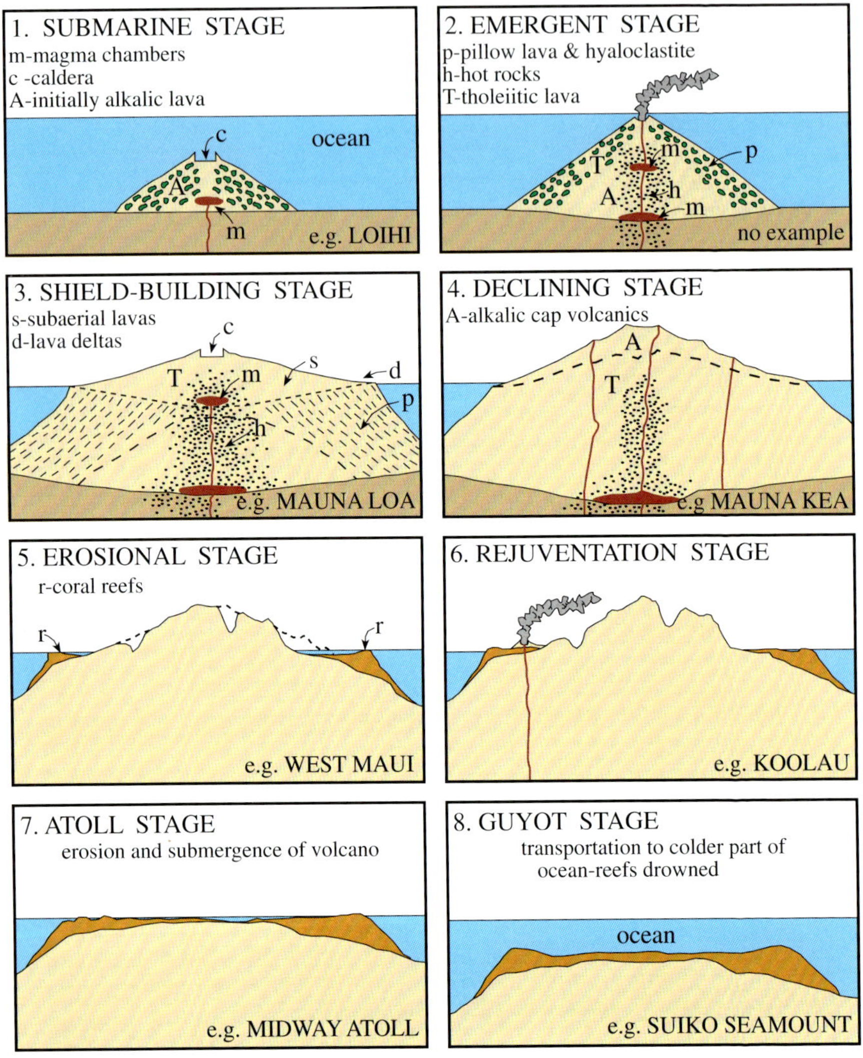

Schematics of **stages in the geologic history** of a typical major volcanic mountain/island in the Hawaiian-Emperor Volcanic Chain. Stage 4 is also referred to as the alkalic-cap or post-shield stage. [modified from Walker (1990), as modified from Stearns (1946), Macdonald et al. (1983) and Peterson and Moore (1987)]

Volcano Building Stages
High Islands of Hawaii

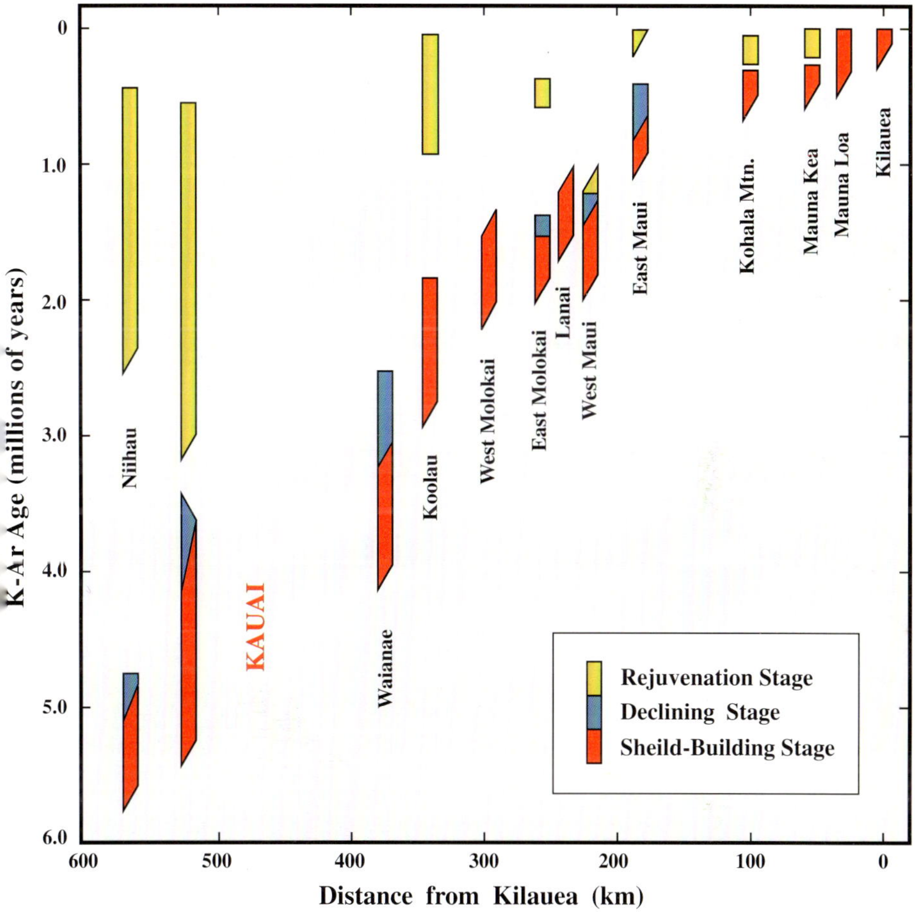

Known durations of the three **major stages of volcano building** for the principal high Hawaiian Islands plotted against the distance from the active Kilauea volcano on the island of Hawaii. Major stages include: **1**) the initial, major, **shield-building stage** (red), **2)** a minor **post-shield (or declining) stage** (blue), and **3)** the final **rejuvenation stage** (yellow). On Kauai the shield stage is represented by the Napali Member of the Waimea Canyon Basalt (formerly known as the Waimea Canyon Volcanic Series), the post-shield stage is represented by the Olokele, Haupu and Makaweli Members of the Waimea Canyon Basalt, and the rejuvenation stage is represented by the Koloa Volcanics (formerly known as the Koloa Volcanic Series). (modified from Clague and Dalrymple, 1989)

GEOLOGY OF KAUAI

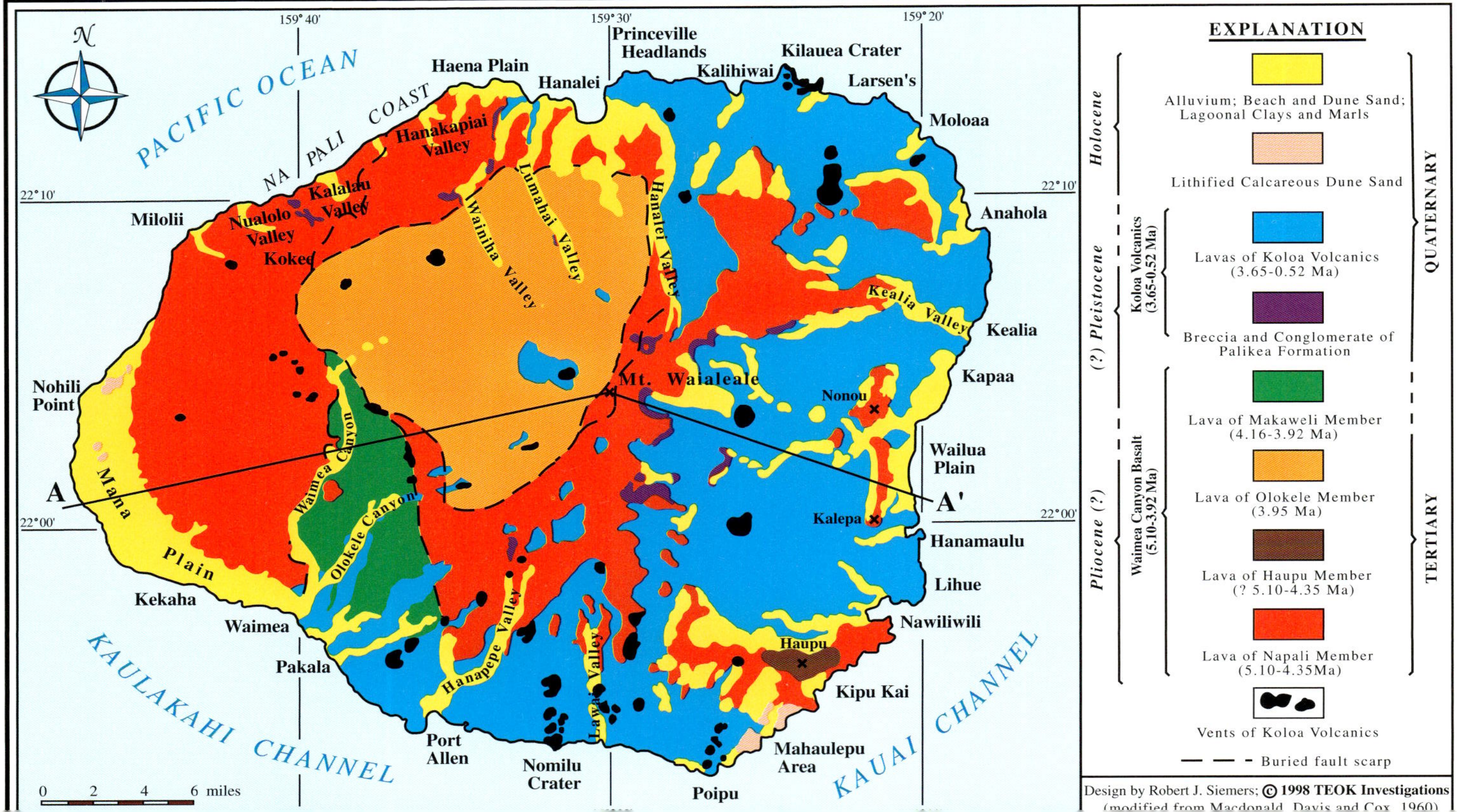

Design by Robert J. Siemers;
(modified from Macdonald, Davis and Cox, 1960)

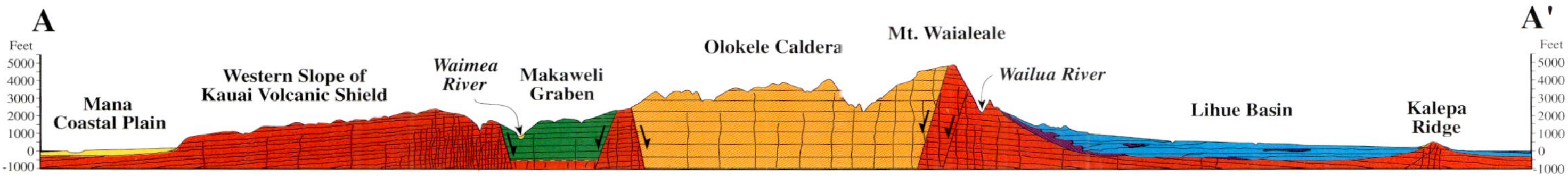

SECTION ALONG A-A'

Geologic map and schematic geologic cross-section of Kauai. Such maps display the distribution, orientation and mutual relationships of major rock types. For a geologist they are absolutely critical displays. This map, published in 1960, still displays the main geologic features of Kauai. Island exposures of the **Napali Member** of the Waimea Canyon Basalt (shown in red) have been dated radiometrically at 4.35-5.10 million years old; however, those rocks represent only the uppermost portion of the basaltic rocks that formed the shield volcanic mountain-island. Note how the lava flows dip gently, at 8-10° away from the center of the island. Similar volcanic rocks comprise the bulk of the mountain to its base, some 18,000 below sea level. Hawaiian volcanoes are built up from the sea floor in as little as 0.5 to 1.5 million years, so the oldest portion of Kauai would only be around 6 million years old. The **Olokele Member** (orange) is composed of somewhat thicker, nearly flat-lying basalt flows thought to represent post-shield, caldera-filling volcanics. Only one radiometric date, at 3.95 my, has ever been reported for those lava flows. The **Makaweli Member** (green), at 4.16-3.92 my, fills a down-dropped fault-like feature known as the Makaweli Graben. The **Haupu Member** (brown) apparently represents lavas that filled a minor flank caldera near the southeastern portion of the present island. **Koloa Volcanics** (blue) represent the rejuvenation stage lava flows and explosive vent deposits that erupted around 3.65-0.52 million years ago. Although they cover most of the eastern portion of the island, they represent only a thin, veneer-like volcanic layer that comprises less than 1% of the bulk of the mountain-island. Sedimentary deposits represented by the **Palikea Formation** (purple) occur mainly between the Waimea Canyon Basalt and the Koloa Volcanics and represent mainly cliff talus and stream sediments eroded and deposited during the brief break in volcanic activity between the shield-building and rejuvenation stages. Finally, note the relatively young **sedimentary deposits** (yellow) of the various river valleys and the Mana Coastal Plain. (Map and cross-section modified from Macdonald, Davis and Cox, 1960).

Kauai Topography

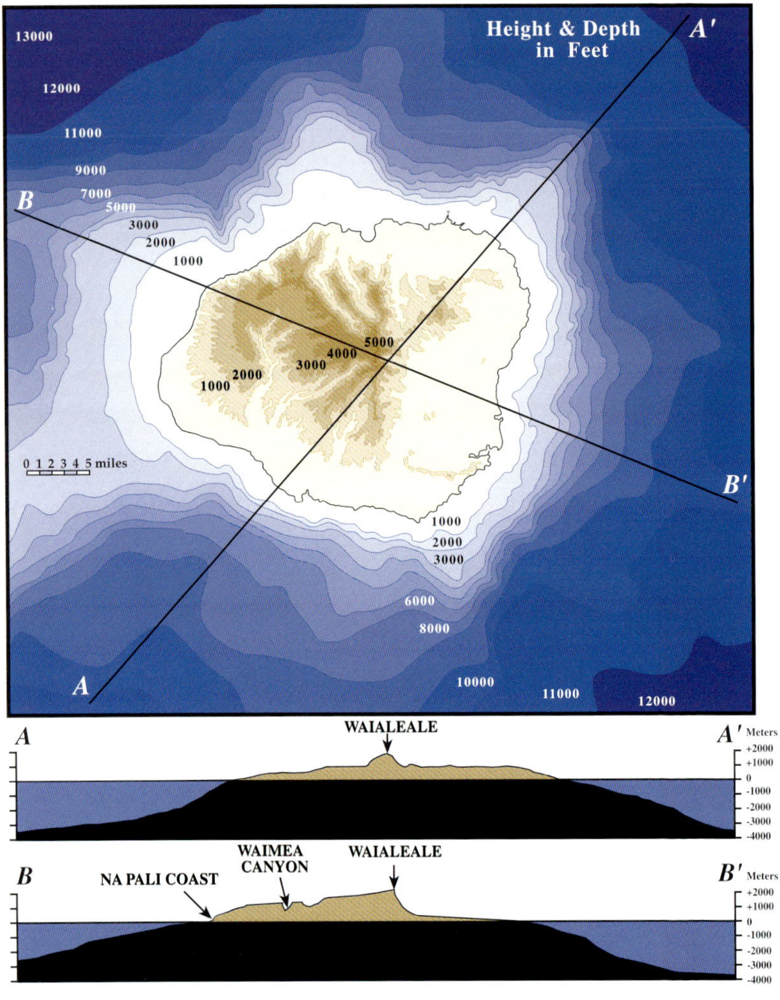

Map and profiles of the **Kauai shield volcanic dome** illustrating major topographic features. The east-west profile B-B' illustrates the **steep eastern scarp of Waialeale** which was formed in part by collapse of the Lihue Basin to the east, and which results in the dramatic **orographic effect** that produces extreme annual rainfall of 400-600 inches on the mountain and across the Alakai Swamp. The **Waimea Canyon** "notch" resulted from a major collapse of the shield dome. The **Na Pali scarp** probably represents a significant structural failure of the volcanic dome in which a major northwestern portion of the island slumped down slope. (modified from Macdonald, Davis and Cox, 1960)

Major Structural Features of Kauai

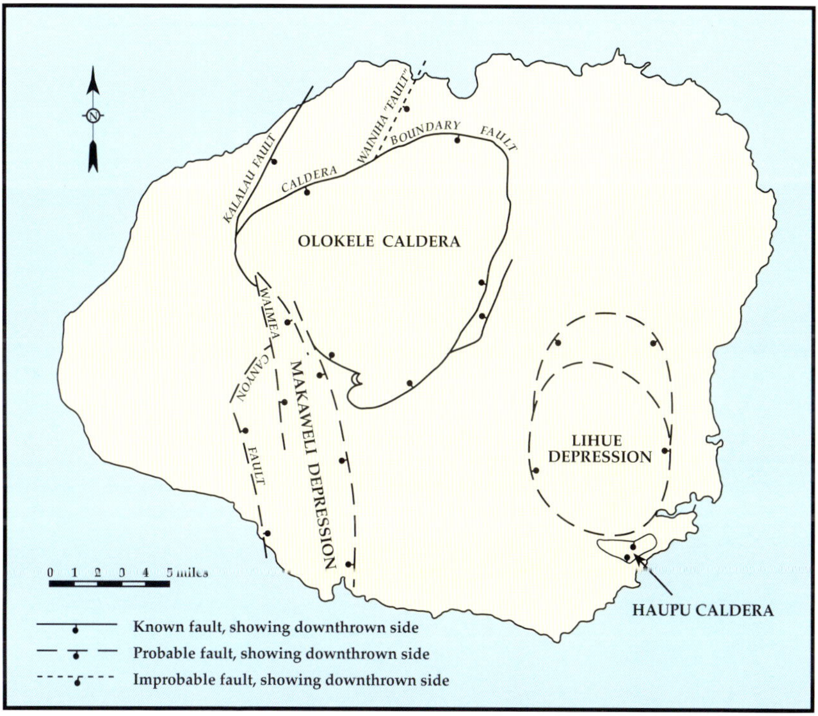

General map of the island of Kauai showing the distribution of **major faults and collapse features**. The original major shield building phase of volcanic dome development on Kauai continued from about 5.1 to 4.35 million years ago. The Olokele Caldera and Makaweli Depression structural features were infilled with post-shield alkalic volcanics of the Olokele Member and Makaweli Member, respectively, of the Waimea Canyon Basalt. Such post-shield volcanic eruptions continued until up to about 3.92 million years ago. Rejuvenation stage volcanics of the Koloa Volcanics (age 3.65-0.52 Ma) extend across the eastern half of the island and partly infill the Lihue Basin depression. (map modified from Macdonald, Davis and Cox, 1960)

Waimea Ca

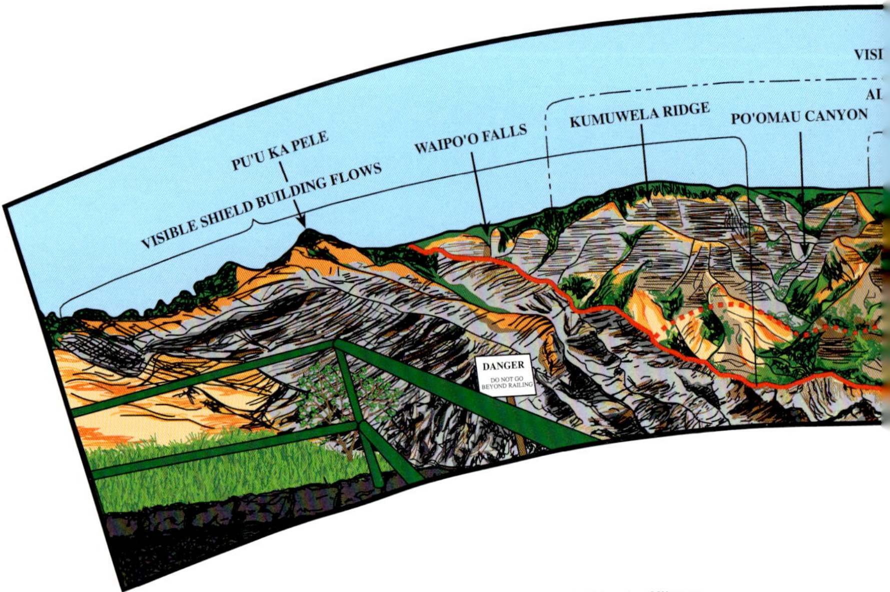

VISI
AL
KUMUWELA RIDGE
PO'OMAU CANYON
WAIPO'O FALLS
PU'U KA PELE
VISIBLE SHIELD BUILDING FLOWS
DANGER
DO NOT GO BEYOND RAILING

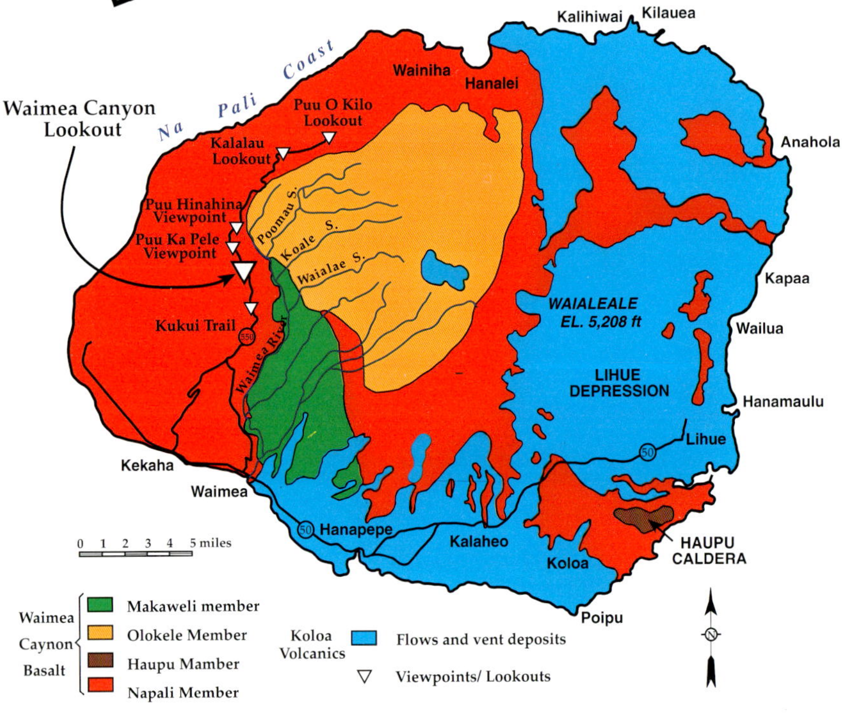

Waimea Canyon Lookout
Na Pali Coast
Kalihiwai
Kilauea
Wainiha
Hanalei
Puu O Kilo Lookout
Kalalau Lookout
Puu Hinahina Viewpoint
Puu Ka Pele Viewpoint
Poomau S.
Koale S.
Waialae S.
Kukui Trail
550
Waimea River
Anahola
Kapaa
Wailua
WAIALEALE EL. 5,208 ft
LIHUE DEPRESSION
Hanamaulu
Lihue
50
Kekaha
Waimea
50
Hanapepe
Kalaheo
Koloa
HAUPU CALDERA
Poipu
0 1 2 3 4 5 miles
Waimea Caynon Basalt
Makaweli member
Olokele Member
Haupu Mamber
Napali Member
Koloa Volcanics
Flows and vent deposits
Viewpoints/ Lookouts
N

on Lookout

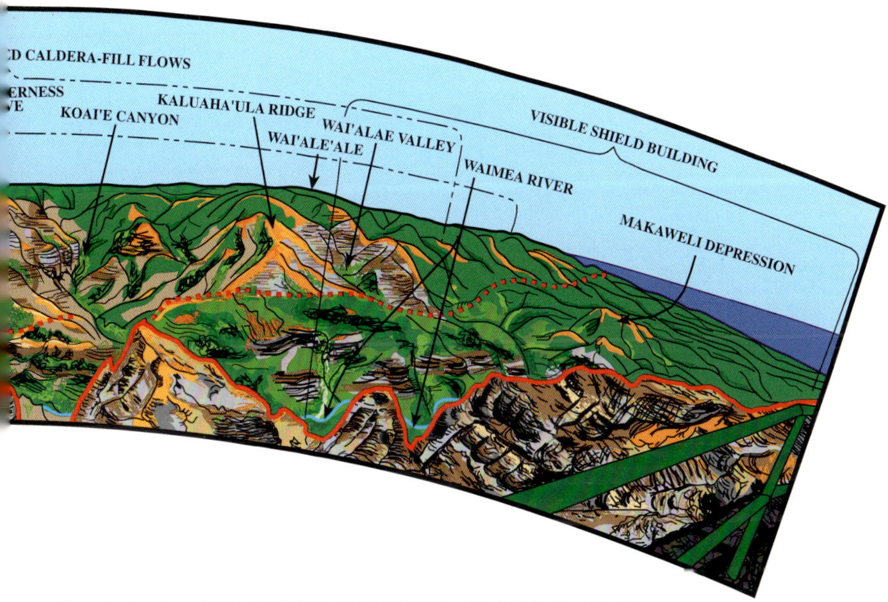

BOUNDARY OF MAKAWELI DEPRESSION ACROSS THE CANYON

INDICATES LIMIT OF SHIELD BUILDING FLOWS ON NEAR SIDE OF CANYON

Panoramic view from the **Waimea Canyon Lookout**, which is located along **State Hwy 550** leading up to **Koke'e State Park** (see simplified geologic map at left showing principal stops along the highway). Shield-building lava flows of the **Napali Member** of the Waimea Canyon Basalt (age **5.1-4.35 m.y.**) are visible to the left and dip 6° to 12° west-northwest away from the center of the volcanic shield that comprises the island of Kauai. The flat-lying, possible caldera-filling lavas of the post-shield **Olokele Member** of the Waimea Canyon Basalt (youngest age of **3.95 m.y.**) are visible across much of the panoramic view. Lavas of the **Makaweli Member** of the Waimea Canyon Basalt (age **4.16-3.92 m.y.**) are present to the south (lower right portion of view) where they fill a down-dropped, structural depression. (panoramic diagram modified from Roelofs, 1993)

K-Ar Dates on Kauai
(Ages in Millions of Years Ago)

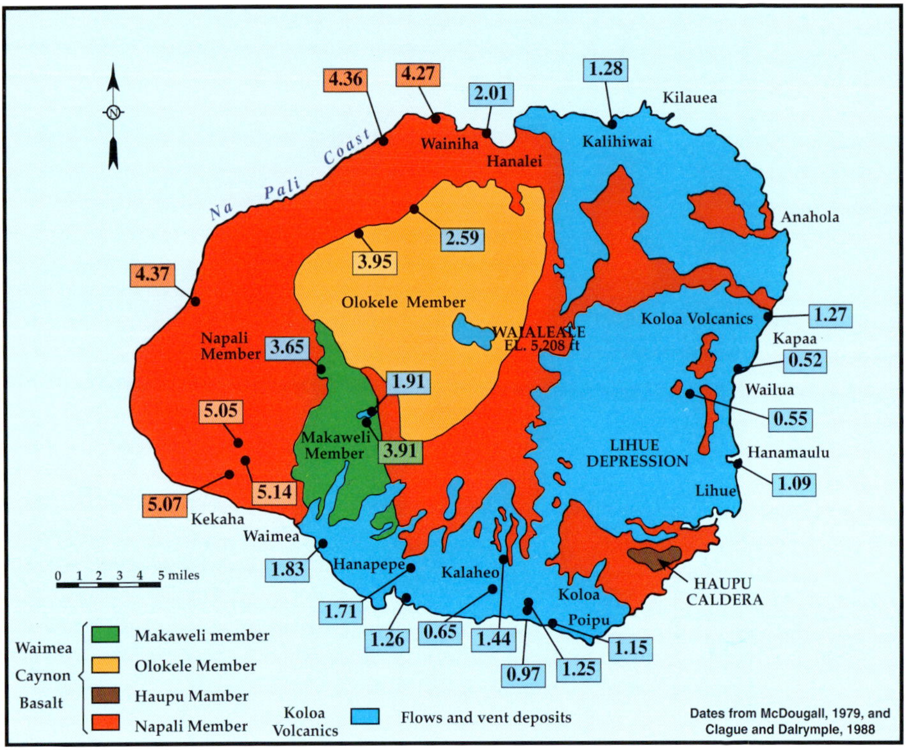

Simplified geologic map of Kauai showing the locations of **radiometric age dates** of the island's major volcanic rock units. The absolute ages, in millions of years, were determined using the **potassium/argon** natural radioactive decay system. Lavas of the **Napali Member** of the Waimea Canyon Basalt represent the prominent shield-building stage of volcanism, which built the mountain/island about **5.1-4.3 million years** ago. Post-shield lavas include the **Olokele Member** (youngest age 3.95 m.y.) which filled a caldera-like structural depression, the **Makaweli Member** (age **4.2-3.9 m.y.**) which filled a north-south trending, graben-like depression, and the **Haupu Member** (precise age not known) which filled a flank caldera at the southeastern part of the island. Lava flows and explosive volcanic vents of the **Koloa Volcanics** cover most of the eastern half of the island and represent the rejuvenation stage of volcanism which occurred **3.6-0.5 million years ago**. In general the Koloa Volcanics become progressively younger to the east; all flows older than 1.7 m.y. occur in the west-northwest half of the island (e.g. west of Hanapepe Valley and Hanalei Valley) and all flows younger than 1.5 m.y. occur in the east-southeastern half (e.g., east of Hanalei and Hanapepe valleys). The youngest lavas dated (0.52-0.55 m.y.) occur in the Wailua Valley area. (modified from McDougall, 1979, and Clague and Dalrymple, 1988)

Explosive Volcanic Vents on Kauai

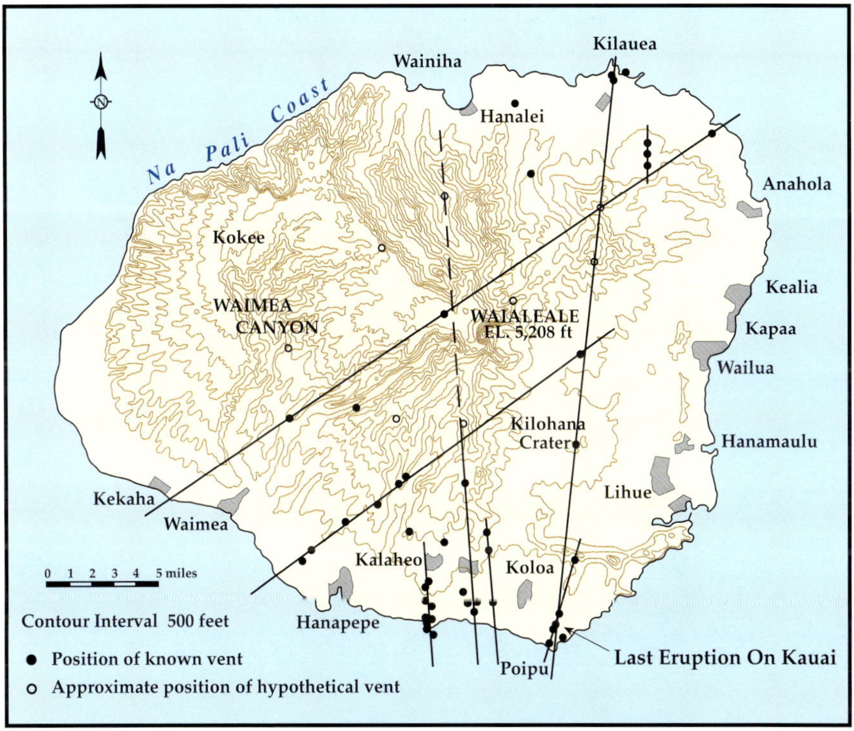

General map of Kauai showing the distribution of **volcanic vents** of the rejuvenation stage **Koloa Volcanics** (age 3.65-0.52 million years). The apparent alignment of vents suggests the presence of large fissure-like fractures in the shield volcanic dome through which the late-stage lavas and explosive volcanics were emitted. These volcanic vents represent the last volcanic events on Kauai; however, absolute geologic ages have not yet been determined for any of these features. (modified from Macdonald, Davis and Cox, 1960)

Geologic Stages of Kauai

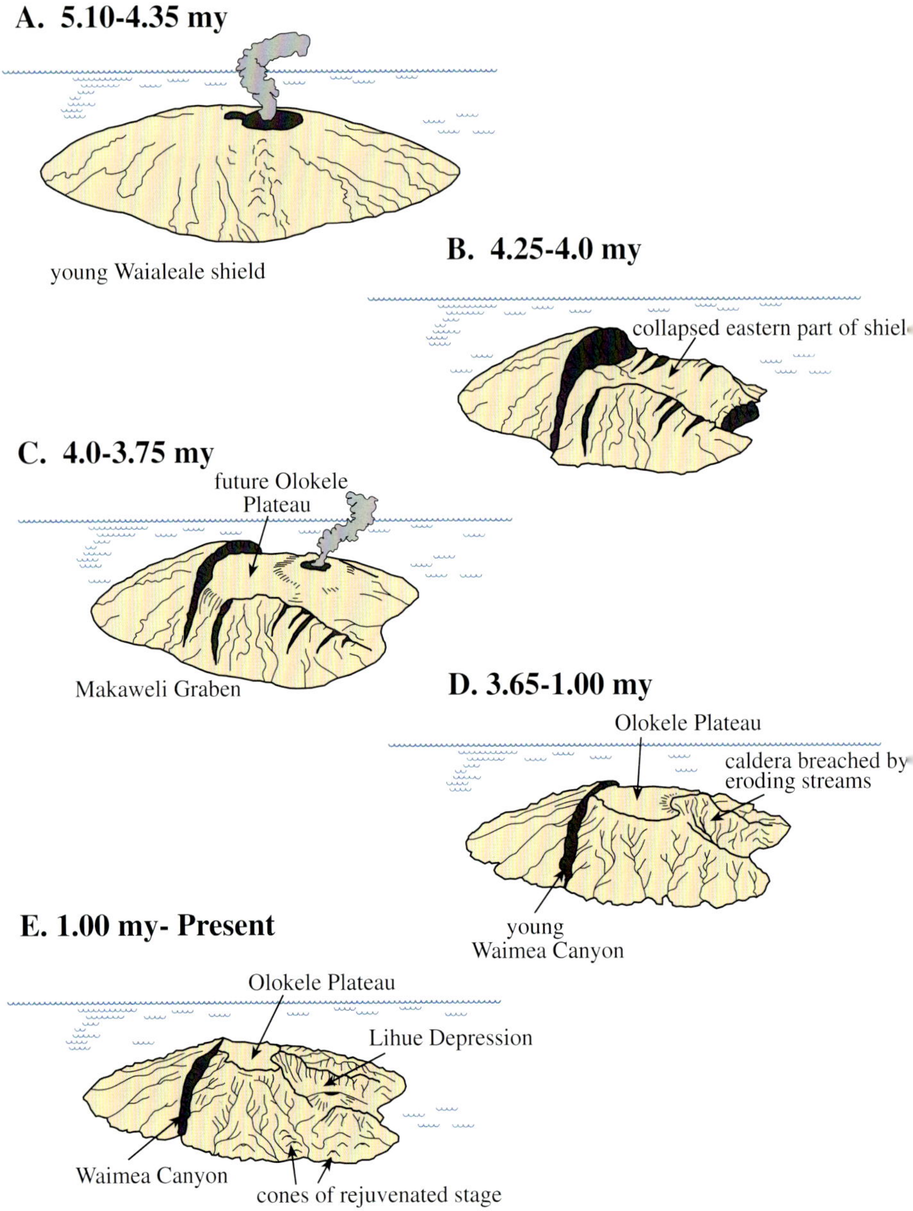

Schematic of various stages in the development of the **island of Kauai**. (modified from Hazlett and Hyndman, 1996)

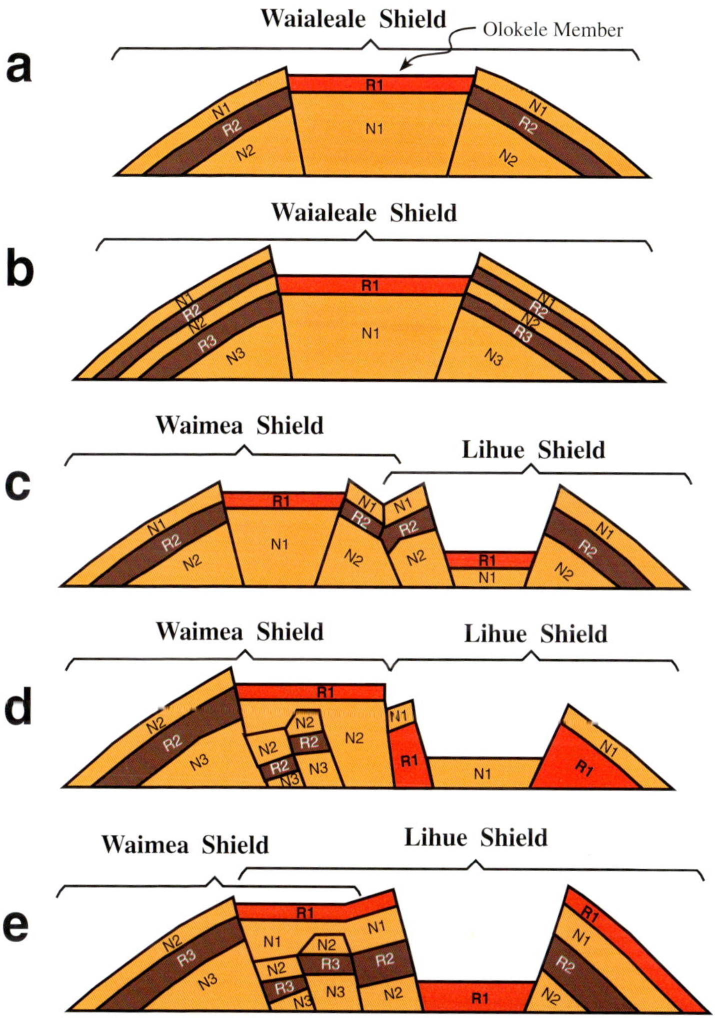

Utilizing a technique know as **magnetostratigraphy** geologists recently have hypothesized that Kauai may actually have been formed by **two shield volcanoes**, rather than just the one Waialeale Shield (a-b). Layers of lava rock formed (i.e., cooled) during **normal** configurations of the Earth's magnetic field, such as it is today, are labeled **"N"**. Layers formed when the Earth's magnetic poles were **reversed** are labeled **"R"**. The **five hypotheses** shown here display what the island would have looked like prior to erosion and to collapse of its eastern portion to form the Lihue Basin. The **Olokele Member** of the Waimea Canyon Basalt, labeled "R1", is shown both as a **caldera fill** of the Waialeale Shield (a-b), or hypothesized Waimea Shield (c-d), and as lava flows formed as part of the **western flank** of a hypothesized Lihue Shield (e). (Modified from Holcomb, et al., 1997)

Kauai Rainfall

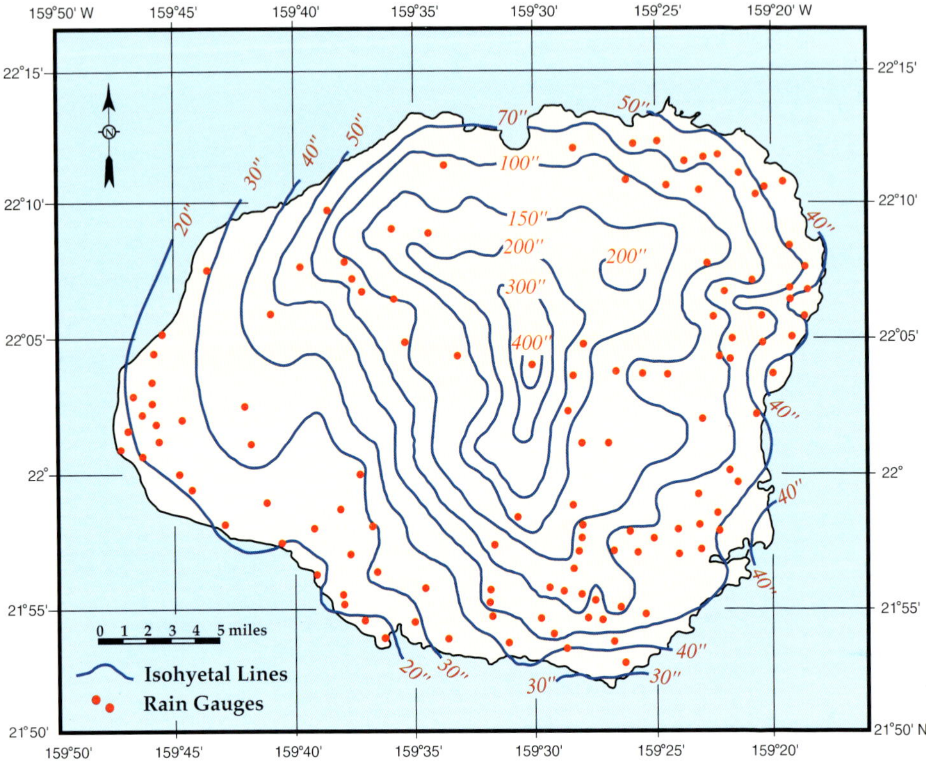

Map of Kauai showing the general distribution of **annual rainfall** (in inches). Isohyetal lines are from the U.S. Weather Bureau. The extreme rainfall at Mt Waialeale is a result of orographic cooling of warm, moist tradewind-driven air as it is forced up the steep eastern scarp of the mountain. (modified from Macdonald, Davis and Cox, 1960)

Relief model of the island of Kauai viewed obliquely from the south. The Puu ka pele slope, west of Waimea Canyon, is underlain by the **Napali Member** of the Waimea Canyon Basalt, which represents the prominent shield-building stage of the mountain/island volcano that emerged above the sea around **5.1 million years ago**. The **Waimea Canyon scarp** probably represents a major collapse at the beginning of the post-shield (or declining) stage (approx. 4.25-4.00 million years ago). Post-shield alkalic volcanics of the **Olokele Member** of the Waimea Canyon Basalt may have infilled a major caldera-like collapse structure to form the present-day broad summit area of Mt. Waialeale and the Alakai Swamp; similar volcanics of the **Makaweli Member** of the Waimea Canyon Basalt also partly infilled the graben-like Makaweli depression. Such post-shield volcanic flows continued until **around 3.92 million years ago**. The major east-west trending Haupu Mountain ridge, between Poipu and Lihue, is composed of the **Haupu Member** of the Waimea Canyon Basalt, and is a probable structural remnant of the original shield-building, and/or post-shield, volcanic stage(s) of the island. After a brief **stage of quiescence** and erosion (lasting only 0.25-0.30 million years), **rejuvenation stage volcanics**, represented by lava flows and explosive volcanics of the **Koloa Volcanics** (age **3.65-0.52 Ma**), covered the eastern portion of the island, including partial infilling of the **Lihue depression**. Very late stage **explosive volcanic vents and cones** of the Koloa Volcanics such as Kilohana Crater, Kilauea Crater to the north, and 35-40 other smaller but similar features are present throughout the eastern portion of the island. The very steep eastern facing scarp of Waialeale, formed in part by collapse of the Lihue Basin, produces a dramatic **orographic effect** by which warm, moisture-laden trade winds from the northeast are forced upward and cooled to generate the condensation that gives Mt. Waialeale the reputation as having one of the highest rainfalls in the world. (from Macdonald, Abbott and Peterson, 1983)

Submarine Geologic Deposit

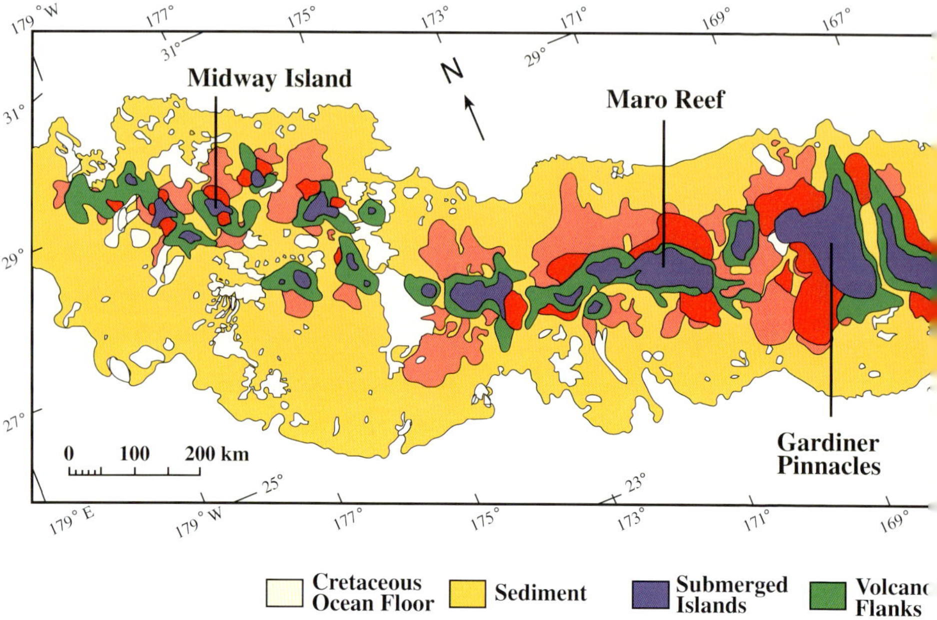

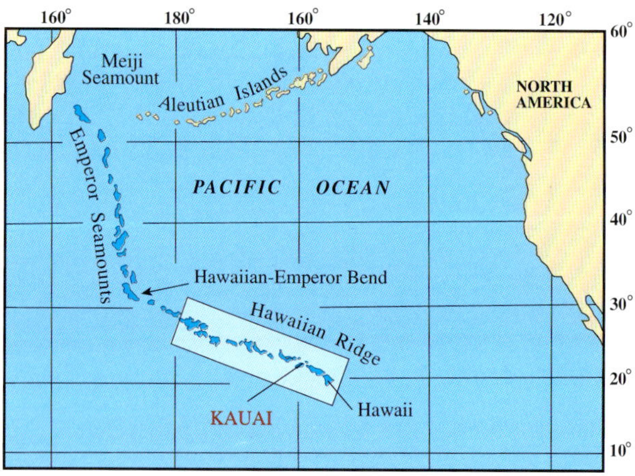

SITE LOCATION MAP

f Hawaiian Volcanic Chain

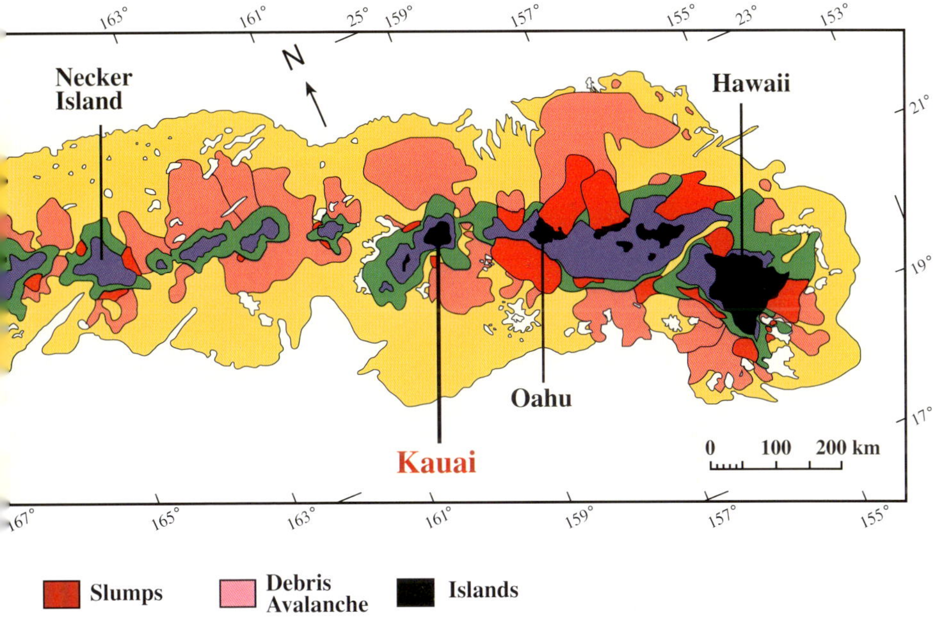

Submarine geologic deposits of the Hawaiian Volcanic Chain from Hawaii to the Midway Islands. Of significance are the **submarine slumps, debris flows and redeposited volcanic/carbonate sediments** indicative of the **destructive processes** that affect ocean volcanic mountain/islands. At least **68 major submarine landslides** more than **20 km** (12.4 mi) **long** have been mapped along the 2200 km (1367 mi) stretch of the Hawaiian Ridge shown here, which developed over the past 30 million years. Geology has been inferred mainly from the long-range side-looking sonar system GLORIA (Geologic Long-Range Inclined Asdic), which records acoustic back-scatter from the seafloor in an effective swath 25-30 km wide. (modified from Moore, Normark and Holcomb, 1994)

Submarine Landslides of Main Hawaiian Islands

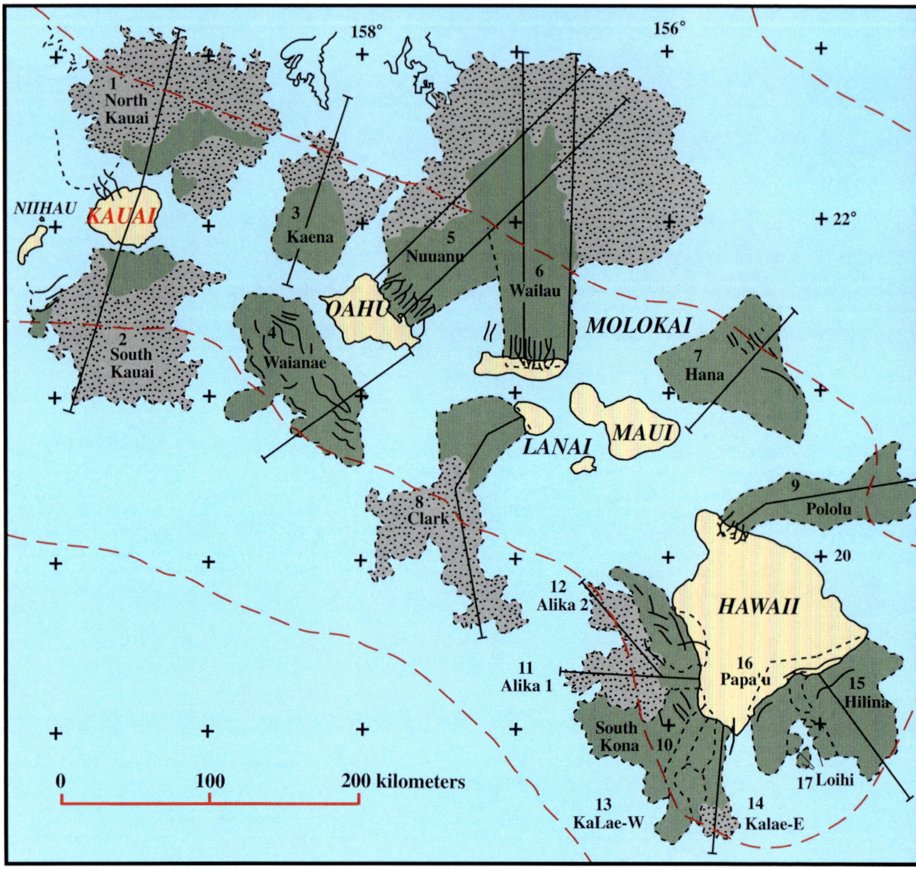

Map of the **southeastern Hawaiian Volcanic Ridge** showing the location of the major volcanic islands and associated **submarine landslides, debris flow deposits** and related features. The brownish-gray stipple represents hummocky areas of probable submarine debris flow deposits. Scarps at the heads of submarine slides are indicated by hachured lines delineating areas of sediment movement (green). Thin downslope lines indicate probable submarine canyons. The axis of the Hawaiian Deep is indicated by interior dashed red line whereas the crest of the Hawaiian Arch is indicated by the outer dashed red line. Locations of longitudinal profiles (shown in following illustration) also are indicated. (modified from Moore, et al., 1989)

Submarine Landslide Profiles of Main Hawaiian Islands

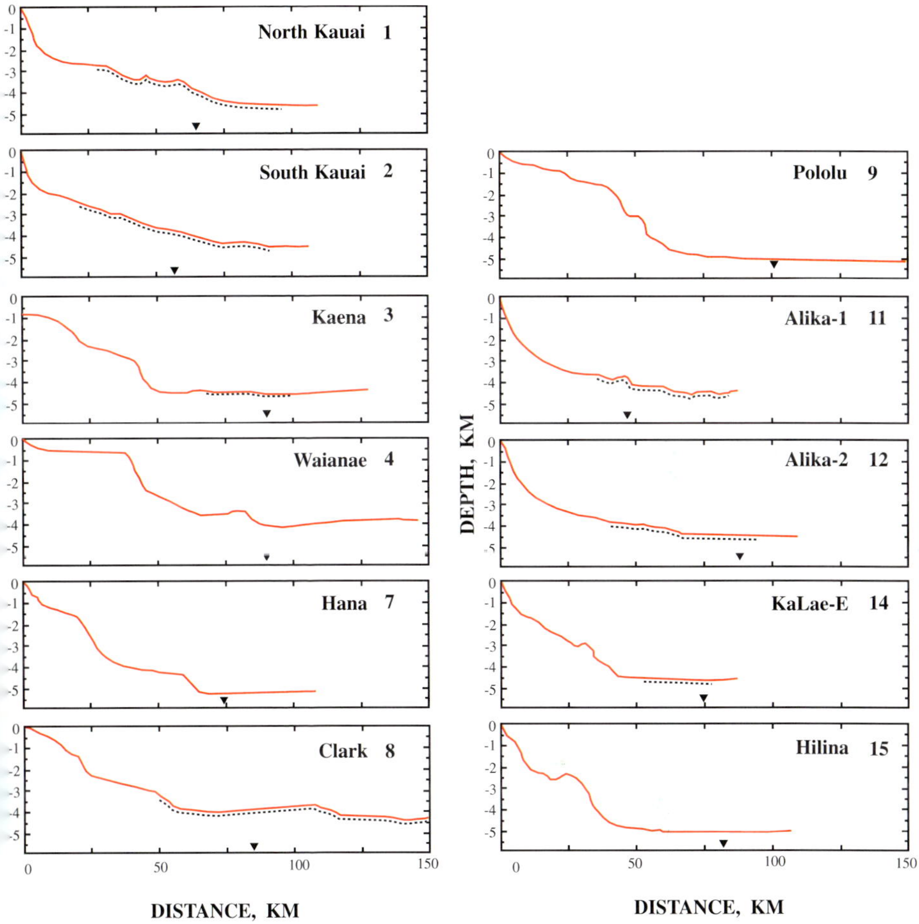

Longitudinal profiles of **submarine landslides** of the major **Hawaiian Islands** at the southeastern end of the Hawaiian Volcanic Ridge. The location of the profiles is indicated in the previous illustration. Hummocky terrain at the top of most slides is indicated by dots. The lower end of the dots marks the limit of the debris avalanche feature. The projected trace of the Hawaiian Deep axis is indicated by the triangle in each profile. Vertical exaggeration is 7.5. (modified from Moore, et al., 1989)

Submarine Features of Kauai and Niihau

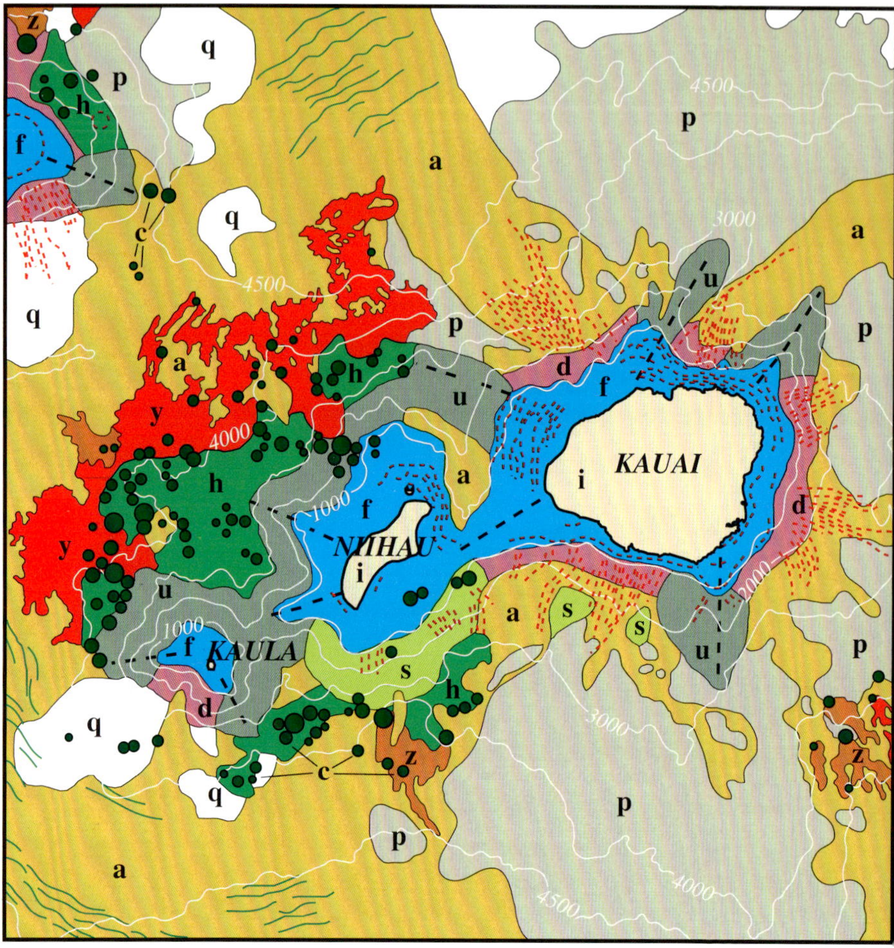

Geologic map of submarine features present around the islands of Kaua'i and Ni'ihau which are indicative of the **destructive phase** of ocean volcanic island history. Map generated from GLORIA **side-scan sonar** data. Features indicated include the following: **i**, present island; **f**, area of former island; **a**, smooth apron of island; **d**, steep submarine flank mantled by debris; **s**, slump terrane with fault blocks; **u**, undisturbed lava flows on island flank; **h**, hummocky volcanic terrane; **p**, coarse avalanche deposits; **q**, fine avalanche deposits; **y** & **z**, mappable lava fields; **c**, small volcanic vent/hill; **radial dashed lines**, submarine canyons and sediment chutes; concentric lines; submerged carbonate reef. (map courtesy of Robin Holcomb, 1996)

CONCLUDING REMARKS

Most people, even those with training in geology, have difficulty dealing with the concept of geologic time. Graphic scenarios never seem to adequately display the vastness of time recorded by the Earth. The Kauai/Earth time line (following page) lists a few events of interest in an attempt to integrate geologic and human history within the 10 orders of magnitude of geologic time.

The geologic and natural history of the present island of Kauai (age greater than 5.1 million years) represents only 0.11% of the time since the origin of the Earth about 4.6 billion years ago. Human history extends to 4-5 million years ago (about the time of the first hominids). The archaeological, anthropological and Hawaiian cultural histories on Kauai occupy only the last 2,000 years, or only about 0.04% of the age of the island. Polynesian and Hawaiian oral history on the island extends to the time of the first occupants. Their written history encompasses less than 220 years, since the time of the first significant contact by Europeans.

The early inhabitants of the isolated volcanic islands of Hawaii were limited in terms of available natural earth materials. Although displaying an advanced culture in terms of language, social structure, craftsmanship and navigational skills, they possessed only stone-age implements at the time of European contact. Hawaiian volcanic rocks do not yield basic metals such as copper, tin and iron, which were so important to the developing cultures of Asia and Europe. Polynesian implements were made of varieties of volcanic rock, limestone (reefrock and beachrock), various woods and plants, bones (including those of humans), shells and the like. Pottery was not produced owing in part to the poor clays generated from the weathering of volcanic rock. Mortar and other types of cement were not used since the Polynesians did not possess that technology, even though limestone is rather plentiful around many of the islands. Obviously this is a good example of how the physical setting affects human cultural development, especially as seen through the eyes of an archaeologist/anthropologist.

Finally, it is interesting to note the changes in the interpretation of the geologic history of the Hawaiian Islands over the last few decades. It might surprise you to know that it was not

KAUAI / EARTH TIME LINE

YEARS BEFORE PRESENT

Kauai / Hawaii event	Years before present	World event
origin of the Earth --	4,600,000,000	
blue-green algae appear --	2,500,000,000	
beginning of Paleozoic (invertebrates) --	570,000,000	
northern end - Emperor Seamounts --	70,000,000	
dogleg bend - Hawaiian/Emperor chain --	43,000,000	
Midway islands, Hawaii --	28,000,000	
oldest volcanic rocks - island of Kauai --	5,100,000	
	4,500,000	**-- early Hominids**
	2,400,000	**-- earliest human stone technology**
	600,000	**-- "Old Stone Age" (Lower Paleolithic)**
youngest volcanic rocks - island of Kauai --	520,000	
Kohala volcanic rocks - island of Hawaii --	430,000	
	250,000	**-- *Homo neanderthalensis***
	100,000	**-- *Homo sapiens***
	30,000	**-- Cro-Magnon man replaces Neanderthals**
sea level in Hawaii -360 to -390 ft --	21,000	
	10,000	**-- end of "New Stone Age" (Neolithic)**
	9,000	**-- Copper Age**
	6,000	**-- Bronze Age**
	3,000	**-- Iron Age**
sea level in Hawaii +5 to +6 feet --	2,000	**-- Polynesian discovery of Hawaii**
	500	**-- European discovery / exploration of Pacific Ocean**
	220	**-- British contact with Hawaii**
Hawaii K-Ar dates; plate tectonics; hot spot theory --	35	
	5	**-- Hurricane Iniki on Kauai**

until the early 1960's, that the geological theories of plate tectonics and magmatic hot spots originated and that the first radiometric dates of the island's volcanic rocks were published. Before that most geologists thought that the Hawaiian Islands had been produced by volcanic eruptions along a long fissure in the oceanic crust. Relative ages of the high volcanic islands, generally with Kauai being the oldest and Hawaii being the youngest, were based on degrees of erosion and the biogeographic distribution of native plants and animals. It is only somewhat recent thought which indicates that Hawaii's volcanic mountain/islands were formed successively as the 3-5 mile thick oceanic crust of the Pacific Tectonic Plate moved northwestward (at a rate of about 3.4 inch/yr) across the Hawaiian Magmatic Hot Spot.

So what can be expected for Kauai? Nihoa is the next Hawaiian island to the northwest. It is an uninhabitable rock less than a mile long and less than 900 ft high, compared with the Garden Island of Kauai which is 33 miles long and over 5,200 ft high. Nihoa's volcanic rocks have been dated at 7.2 million years, compared with Kauai at 5.1 million years. Assuming that Nihoa was at least 5,000 ft high 2.1 million years ago, Kauai would have to subside and be eroded at a rate of 2,050 ft/million years (or at about 1 inch every 40 years) to reach that level. Nihoa is 162 miles northwest of Kauai. With the Pacific Tectonic Plate moving at 3.4 inch/yr it would take about 3,035,000 years for Kauai to move to Nihoa's present location. Using such scenarios, one can estimate that it will take something around 2 to 3 million years for Kauai to be transformed from one of the Earth's most beautiful islands to one preferred mainly by sea birds.

GLOSSARY

Absolute Age. The geologic age of rocks and related geologic features or events given in units of time, usually years, rather than in relation to (i.e., "younger" or "older" than) other geologic features. In Hawaii the absolute age of volcanic rocks is determined by radiometric dating (see radiometric dates).

Alkalic Lava/Basalt. A fine-grained (crystalline) dark-colored volcanic rock poor in silica and relatively rich in minerals with elements such as sodium and potassium. In Hawaii post-shield and rejuvenation stage lavas are mainly alkalic basalt with plagioclase feldspar, calcium-rich augite and olivine.

Aragonite. A calcium carbonate (CaCO3) mineral with orthorhombic crystal structure. It has a greater density and hardness, and less distinct cleavage than calcite (also composed of CaCO3) and is less stable than calcite at Earth's surface temperature and pressure. Organisms such as corals and molluscs (both pelecypods and gastropods) commonly produce skeletal material composed of aragonite, which eventually converts to calcite.

Atoll. In general, a ring-shaped coral/algal reef appearing as a low, roughly circular to elliptical island with a central lagoon and surrounding open ocean. Such low, reef islands are common in the central and western Pacific Ocean. Northwest of Kauai atolls have formed as high volcanic islands have subsided and been replaced by the fringing reefs that continued their upward growth.

Basalt. A fine-grained (crystalline) dark- to medium-colored (i.e., gray to black) igneous rock formed when molten lava is extruded onto the Earth's surface and cools rapidly. When erupted from volcanoes, basaltic lava tends to be fluid and floods evenly over large areas. Basalt covers about 70% of the Earth's surface as it is the main rock type forming the ocean floor. The Hawaiian Islands consist primarily of basalt.

Bathymetry. The measurement of water depths in the ocean and charting of the topography of the ocean floor.

Calcite. A rock-forming mineral composed of calcium carbonate (CaCO3), with trigonal crystal structure and perfect rhombohedral cleavage. Calcite is more stable at Earth's surface than aragonite (also composed of CaCO3). Most coralline algae, which is a principle constituent of coral/algal reefs, and organisms like foraminifers and echinoderms produce skeletal material composed of calcite. Uplifted reef platforms and beachrock commonly are cemented by calcite.

Caldera. A large (1 to >10 km) roughly circular volcanic crater formed by the collapse of the central portion of a volcano as the magma chamber below empties. Calderas are characterized by steep sides and a flat bottom like those of Kilauea Crater on the island of Hawaii.

Clay/Claystone. Clay is both a textural (e.g., grain-size) term and a compositional term. Clay particles are less than 4 microns (0.004 mm) in diameter; a claystone is a rock composed mainly of clay-size particles. A group of minerals known as "clay minerals" includes minerals like kaolinite, illite, and montmorillonite, which also commonly are clay size. Soils contain abundant clay-size particles and, at the same time, clay minerals.

Coral/Algal Reef. A ridge- or mound-like organic structure built mainly by corals and coralline algae. In Hawaii, Tertiary to Holocene, and modern, reefs are generally referred to as "coral" reefs but may actually be composed mainly of coralline algae, which encrusts and binds the structure.

Conglomerate. A coarse-grained sedimentary rock composed of rounded, gravel-size grains (i.e., pebbles, cobbles and boulders; diameter >2 mm) which have been cemented, and/or pressed together to form a rock. In Hawaii such sedimentary deposits are common in mountain streams and along high-energy coasts.

Debris Avalanche/Flow. A rapid (avalanche) or slow (flow) mass movement of material (commonly water-saturated) downslope in response to gravity. In Hawaii such destructive events occur both above (i.e., subaerial) and below (i.e., subaqueous) sea level owing to the destruction of the islands by the atmosphere and hydrosphere.

Dike. A cross-cutting (discordant) tabular (or planar) rock feature, usually an igneous intrusive, which cuts at relatively high angles across surrounding planar rocks (e.g., older lava flows).

Dolomite. A rock forming mineral composed of calcium-magnesium carbonate, MgCa(CO3)2, and with trigonal crystal structure. Dolomite forms by replacement of calcite and/or aragonite where magnesium is abundant.

Fault. A fracture or break in a series of rocks along which movement has taken place. The movement may vary from a few centimeters to a few kilometers.

Fracture. A general term for any break in a rock, whether or not it shows displacement, owing to mechanical failure. Fractures include cracks, joints and faults.

GLORIA. Acronym for "Geologic Long-Range Inclined Asdic", a side-looking sonar system for recording acoustic back-scattered energy from the sea-floor. Such a system recently has been used successively to examine the topography of the deep seafloor around the Hawaiian Islands.

Graben. A structural feature where an elongate block of the earth's crust, bordered by normal faults, has been down-dropped, or lowered, relative to the surrounding rocks.

High Island. In the Pacific Ocean a volcanic mountain/island for which the island portion still extends well above sea level (usually at least several hundred feet). The eight main Hawaiian Islands of Hawaii, Maui, Molokai, Lanai, Kahoolawe, Oahu, Kauai and Niihau are excellent examples of high islands.

Hot Spot (Magmatic). A stationary cylindrical/cone-shaped magma chamber within the upper mantle of the Earth that intermittently pushes magma upward through the crust as the plate moves across the chamber. The Hawaiian Island Magmatic Hot Spot is probably the Earth's best example of such a feature.

Isohyetal Lines. Contour lines on a map connecting points of equal rainfall.

Limestone. A sedimentary rock consisting chiefly of calcium carbonate, primarily in the form of the mineral calcite. Geologically young limestones around the Hawaiian Islands (e.g., coral/algal reefs, beachrock) may also contain significant amounts of the calcium carbonate mineral aragonite.

Low Island. In the Pacific Ocean a relatively low relief coral/algal reef atoll, bank, shoal, etc. composed of calcium carbonate (CaCO3) organic deposits. In contrast to volcanic high islands, they usually extend only a few feet, or tens of feet above sea level. Many of the Hawaiian Islands between Kauai and Midway Island are low islands.

Magma. Naturally occurring molten, mobile rock material generated at high temperature and pressure in the lower crust and upper mantle of the Earth. Such material is capable of intrusion into the crust or extrusion onto the Earth's surface (both below and above sea level). Below Hawaii magma is formed by the melting of upper mantle crystalline rocks at 60-170 km (35-105 mi) depth.

Magnetostratigraphy. Study of layered rocks in which different layers record changes in the magnetic field of the earth, e.g., "normal" and "reversed" magnetic orientations. In Hawaii large sections of layered lava may display such reversals of the magnetic orientation of iron-rich minerals within the basalt rocks.

Mass Wasting. A general term for the loosening and downslope, gravity-driven movement of soil and rock; e.g., water-saturated landslides, debris avalanches. The Hawaiian volcanic islands have undergone extensive destruction by both subaerial (above sea level) and submarine mass wasting of their slopes in response to interaction with the atmosphere and hydrosphere.

Mud/Mudstone. Fine-grained sedimentary material/or rock, of any composition, with grain size of less than 62 microns (0.062 mm); includes silt and clay size material. Soils are composed mainly of mud-size material.

Mg-Calcite. A variety of the mineral calcite [(Ca,Mg)CO3] with randomly substituted magnesium carbonate in a disordered lattice of calcite. Low magnesium-calcite contains <4% MgCO3, whereas high magnesium-calcite contains 4-19% MgCO3. Some skeletal material (e.g., echinoids and some molluscs) and marine cements are composed of high-magnesium calcite.

Oceanic Mountain/Island. Mountain (usually volcanic) within the ocean basin, the top of which extends above sea level. The eight main islands of Hawaii are excellent examples of ocean volcanic mountain/islands.

Olivine. A green mineral composed of (Mg,Fe)2SiO4, which is common in basic rocks such as basalt. Olivine crystals are common in some of the basaltic lavas of the Hawaiian Islands; they commonly weather out and are concentrated in some beach sands, which then take on a greenish appearance.

Orographic Effect. Rainfall that results when moisture-laden air is forced upward over a high barrier into a cooler altitude causing the moist air to condense into rain droplets. Such is the case in the Hawaiian Islands where moist, tradewind-driven air is forced up and over the high volcanic mountain/islands.

Pillow Lava/Basalt. A general term used for lavas displaying pillow-shaped masses. Such shapes result from the rapid cooling of molten lavas when they encounter water; they range in size from a few inches to a few feet in diameter. Pillow lavas are common in Hawaii where mountain/islands were formed by extrusion of lava directly on to the sea floor and by lava flowing down the island and into the ocean.

Post-Shield Stage. The stage in volcano building that occurs after the main shield-building stage and commonly includes partial collapse/subsidence (e.g., caldera formation) and minor eruptions that partially fill such collapse structures.

Radiometric Dating. A method of calculating an absolute age (in years) for geologic materials by measuring the presence of radioactive elements and their decay products; e.g., potassium-40/argon-40. Such elements have known decay rates so the relative amounts of original material and decay products is related to the amount of time the material has been decaying. In Hawaii the absolute ages of volcanic materials have been determined by radiometric dating.

Rejuvenation Stage. Relatively late stage of development in the life of an oceanic mountain/island volcano in which lavas and ash are extruded through fissures and vents. The last stage of extrusion commonly involves explosive eruptions that form cinder cones. On Kauai the rejuvenation stage of volcano building continued until about 500,000 years ago and included both lava flows and explosive eruptions from as may as 40 volcanic vents.

Relative Age. The geologic age of a rock, or geologic feature/event only in relation to other rocks, features or events. Therefore, the ages of rocks and geologic features or events can be referred to only as "older" or "younger", and not in terms of absolute time, such as years.

Sand/Sandstone. Sedimentary material/rock composed of grains (of any mineral or chemical composition) with diameter between silt (62 microns; 0.062 mm) and granule (2000 microns; 2 mm). Most beaches, and associated wind-blown dune deposits in Hawaii are composed of sand size sediment.

Seamount. Volcanic mountain/island whose top is below the surface of the ocean owing to subsidence (i.e., sinking) and/or erosion. All of the previous volcanic mountain/islands of the Emperor Volcanic Chain and some of the volcanic mountain/islands of the Hawaiian Volcanic Chain have been eroded and sunk below sea level to form seamounts.

Shield-Building Stage. Initial stage of volcanic mountain/island in which the main body of the mountain is built by volcanic eruptions both below and above sea level. The rounded profile of Hawaiian volcanic mountain/islands displays the shape of a shield such as that used by Roman warriors.

Sill. A tabular (or planar) rock feature, usually an igneous intrusive, that parallels (or is concordant with) the planar structure of the surrounding rock (e.g., older lava flow layers).

Subsidence. A local mass movement which involves the sinking, or settling, of a large portion of the Earth's crust relative to its surrounding parts. All of the volcanic mountain/islands of the Hawaiian Islands have undergone various amounts of subsidence, some greater than 10,000 feet.

Tectonic Plate. Large portion of Earth's crust (i.e., oceanic or/and continental) which moves in response to lateral forces in the upper mantle. The extensive chain of volcanic mountain/islands, atolls, seamounts, etc. of the Hawaiian-Emperor Volcanic Chain formed as the Pacific Tectonic Plate moved northward and northwestward across the Hawaiian Magmatic Hot Spot in the Earth's mantle.

Tholeiitic Lava/Basalt. Finely-crystalline volcanic rock composed primarily of plagioclase, pyroxene (augite) and iron oxide minerals; with little or no olivine. Such volcanic rocks comprise most of the Hawaiian volcanic mountain/island masses.

Volcanic Chain. A linear cluster of volcanic mountain/islands; e.g., Hawaiian-Emperor Volcanic Chain.

REFERENCES CITED

BRADSHAW, M., and WEAVER, R., 1993, Physical geography; an introduction to earth environments: Mosby, St Louis, 640 pp.

CARSON, H.L., and CLAGUE, D.A., 1995, Geology and Biogeography of the Hawaiian Islands. In Wagner, W. L., and Funk, V. A., eds., Hawaiian Biogeography; Evolution on a Hot Spot: Smithsonian Institution Press, Wash., DC, p. 14-29.

CLAGUE, D.A., and BOHRSON, W., 1987, Geologic field guide to Waimea Canyon, Kauai, Hawaii: Geological Society of America Centennial Field Guide -- Cordilleran Section, 1987, p. 1-4.

CLAGUE, D.A., and DALRYMPLE, G.B., 1987, The Hawaiian-Emperor volcanic chain. In Decker, R.W., Wright, T. L., and Stauffer, P.H., eds., Volcanism in Hawaii: U.S. Geological Survey Professional Paper 1350, U.S. Government Printing Office, Wash., DC, p. 1-54.

CLAGUE, D.A., and DALRYMPLE, G.B., 1988, Age and petrology of alkalic postshield and rejuvenated-shield lava from Kauai, Hawaii: Contrib. Mineral Petrol, v. 99, p. 202-218.

CLAGUE, D.A., and DALRYMPLE, G.B., 1989, Tectonics, geochronology and origin of the Hawaiian-Emperor volcanic chain. In Winterer, E.L., Hussong, D.M., and Decker, R.W., eds., The Geology of North America; Volume N, The Eastern Pacific Ocean and Hawaii: The Geological Society of America, Boulder, CO, p. 188-217.

HAZLETT, R.W., and HYNDMAN, D.W., 1996, Roadside Geology of Hawaii: Mountain Press Pub. Co., Missoula, 307 pp.

HOLCOMB, R.T., REINERS, P.T., NELSON, B.K., and NUNI-LYN, E.S., 1997, Evidence for two shield volcanoes exposed on the island of Kaua'i, Hawaii: Geology, v. 25, no.9, p. 811-814.

KLEIN, F.W., and KOYANAGI, R.Y., 1988, The seismicity and tectonics of Hawaii. In Winter, E.L., Hussong, D.M. and Decker, R.W., eds., The Geology of North America; Volume N, The Eastern Pacific Ocean and Hawaii: The Geological Society of America, Boulder, CO, p. 238-252.

McDOUGALL, I., 1979, Age of shield-building volcanism of Kauai and linear migration of volcanism in the Hawaiian Island chain: Earth and Planetary Science Letters, v. 46, p. 31-42.

MACDONALD, G.A., ABBOTT, A.T., and PETERSON, F.L., 1983, Volcanoes in the Sea: The Geology of Hawaii, 2nd edition: Univ. Hawaii Press, Honolulu, 517 pp.

MACDONALD, G.A., DAVIS, D.A., and COX, D.C., 1960, Geology and groundwater resources of the island of Kauai, Hawaii: Hawaii Division of Hydrography, Bull. 13, 212 pp, colored geologic map.

MOORE, J.G., and CLAGUE, D.A., 1992, Volcano growth and evolution of the island of Hawaii: Bulletin of Geological Society of America, v. 104, p. 1471-1484.

MOORE, J.G., CLAGUE, D.A., HOLCOMB, R.T., LIPMAN, P.W., NORMARK, W.E., and TORRESAN, M.E., 1989, Prodigious submarine landslides on the Hawaiian Ridge: Journal of Geophysical Research, v. 94, p. 17,465-17,484.

MOORE, J.G., NORMARK, W.R., and HOLCOMB, R.T., 1994, Giant Hawaiian underwater landslides: Science, v. 264, p. 46-47.

MOORE, J.G., NORMARK, W.R., and HOLCOMB, R.T., 1994, Giant Hawaiian Landslides: Annual Review Earth and Planetary Sciences, v. 22, pp. 119-144.

PETERSON, D.W., and MOORE, R.B., 1987, Volcanism in Hawaii. Geologic history and evolution of geologic concepts, island of Hawaii: U.S. Geological Survey Professional Paper 1350, p. 149-189.

ROELOFS, F., 1993, Canyon Country Excursions: Waimea Canyon and Kaluapuhi Trail: Moanalua Gardens Foundation, Honolulu, 32 pp.

STEARNS, H.T., 1946, Geology and groundwater resources of the island of Hawaii: Hawaii Division of Hydrography, Bull. 9, 363 pp., colored geologic map.

WALKER, G.P.L., 1990, Geology and volcanology of the Hawaiian Islands: Pacific Science, v. 44, p. 315-347.